Animals of the Arctic
the ecology of the Far North

Animals of the Arctic
the ecology of the Far North
by Bernard Stonehouse

Holt, Rinehart & Winston
NEW YORK, CHICAGO, SAN FRANCISCO

BREUMER

The author gratefully acknowledges his reference to
Geography of the Northlands, by Kimble & Good,
published by the American Geographical Society, New York,
in preparation of the maps in this book.

First published in England by Ward Lock Limited

Designed and produced for the publisher
by Eurobook Limited, London

Copyright © 1971 by Eurobook Limited

Published simultaneously in Canada by Holt, Rinehart
and Winston of Canada, Limited

ISBN: 0-03-086699-5

Library of Congress Catalog Card Number: 74-162304

Manufactured in England by
Sir Joseph Causton & Sons Limited, Eastleigh, Hampshire

First American Edition

Introduction 11

12 **Arctic and Subarctic: North Polar Regions**
Polar and subpolar seas · The northlands · Arctic plants and animals

34 **The Changing Polar Climate**
Solar radiation · Formation of the ice caps · Milankovich cycles · Animals and the ice ages

50 **Arctic Seas: the Chain of Life**
Plankton · The sea bed fauna · Whales · Seals · Polar bears

72 **Sea Birds of the Arctic**
Petrels · Phalaropes · Gulls, terns and skuas · Alcids · Divers · Sea ducks

94 **Life on the Tundra**
Arctic soils · Tundra plant communities · Invertebrate animals · Tundra herbivores
Small herbivores: cycle of abundance · Tundra carnivores · Birds of the tundra

152 **Polar Challenge: Man in the Arctic**
Native populations · Exploitation · Marine exploitation · Arctic animals and man

Bibliography 166   *Glossary* 167   *Index* 169

Introduction

Far to the north of cities and civilization, surrounding the north geographical pole, lies an ice-covered, roughly rectangular ocean. Flanked by the northern coasts of Canada, Greenland and Eurasia, it is the centre of a vaguely defined region of persistent cold—the Arctic. "Arctic" first described the northern lands which lay, far from Mediterranean cultures, under the constellation of the Great Bear (Arktos) or Big Dipper. Today the word implies the farthest regions of the north—the polar ocean, bare tundra lands surrounding it, and the northern edge of the boreal coniferous forest. There is no definite geographical Arctic. There is an Arctic Ocean or Sea, but no "Arcticland" or "Arctica" to match the great southern continent of Antarctica.

Administrators avoid the word, especially if they are trying to settle people in the north. In several languages Arctic is synonymous with hostile cold—the bitter cold of hard blue eyes and frigid social gatherings. It is not a word to attract immigrants. Certainly the Arctic is bare, and it offers less comfort and stimulation than most men feel they need. But the simplicity of polar life has a lasting appeal, and those who know the Arctic have found plenty to interest them. Vilhjalmur Stefansson, who spent much of his life in the far north, called it "the Friendly Arctic". Warm human cultures have flourished there for over ten thousand years.

Animals too find it exciting. With all the world to choose from, millions of birds fly there every year; for them it is a lively place to rear a family. Caribou trot over a thousand miles each year to visit the Arctic. Lemmings, voles and snowy owls flourish there; musk-oxen and Arctic hares live as far north as they can. What is the attraction of this curious region of the Great Bear?

Naturalists find the Arctic exciting for many reasons. It is remote, a place where you can get away to watch and think quietly. It is cold, and constant cold is a challenge to living creatures: how plants and animals survive in constant cold is a biological problem with many facets. There is an appealing biological economy about Arctic organisms. The few hundred plants and animals which live in the far north have simple relationships with each other and with the environment. There is none of the rag-bag clutter of the tropics, where thousands of species jostle for space and ecology is a tangled net. Simplicity makes it a good training ground for ecologists; the Arctic is the treeless land where, for once, we can see both the wood and the trees. For plants and animals it is a hard environment but a rewarding one; many forms of living organism have adapted successfully to it, and need no other inducement to live there.

Finally the Arctic is beautiful. It has splendid animals in a dramatic setting, and neither owes anything to man. Primitive man found his niche and became a part of it. Civilized man has shot and bulldozed his way into it, and is destroying it piecemeal for one empty reason after another. There will soon be no part of it untouched by his greed and pollution. Meanwhile the Arctic is there for us to enjoy. I hope that this book will help to spread the enjoyment.

Bernard Stonehouse
Forgandenny
Perthshire

Arctic and Subarctic

North Polar Regions

Svalbard. These islands, forested during warm spells of the glacial period, were more recently capped almost completely by ice.

Previous page:
Ice crystals grow in the frosty air. Shifting ice floes on the Kolyma River, eastern Siberia.

Arctic and Subarctic

The Arctic is the cold region surrounding the North Pole; the Subarctic is the broad ring of land and sea surrounding the Arctic. Where does one begin and the other end?

The boundary used by geographers and international lawyers is the Arctic Circle (Lat. 66° 33′N.). This is the parallel of latitude beyond which the sun does not appear above the horizon across midwinter. Its relation to the earth's angle of inclination and its significance as a solar boundary are shown on page 37. North of the Circle, the sun is absent for longer or shorter periods on either side of midwinter day, and present above the horizon for similar periods across midsummer. This rule is varied slightly by atmospheric refraction, which appears to raise the sun above the horizon when it is actually below.

The Arctic Circle gives a workable definition of the Arctic for administrative or legal purposes, but has no meaning for animals or plants; it is not a good biological boundary. Though living organisms respond to sunlight, the presence or absence of weak winter sunshine hardly affects them. Nor does it affect climate, or any other factor which might influence them directly. There is nothing on the ground to show where the Circle crosses; it traverses tundra, forest and steppe indifferently, and no animal but man knows anything about it.

The boundary which is used in this book is a climatological one with biological implications. In his system of climatic classification Wladimir Köppen defined as a polar climate one in which the mean temperature of the warmest month does not exceed 10°C. (50°F.). In practice, this places the 10°C. July isotherm as the southern limit of the polar region. This definition makes good biological sense, because Köppen noted that the isotherm follows fairly closely a biological boundary—the *tree line*, or northern limit of tree growth. These lines are shown on page 24. So Köppen's Arctic is the northland area where trees will not grow. This fits well with most people's concept of the Arctic, and we can use either the isotherm or the tree line as an Arctic-Subarctic boundary.

BREUMMER

Ice-worn rocks of the Canadian Shield form a natural rockery for tundra herbs. Barren Ground tundra, near Bathurst Inlet, northern Canada.

We need not worry that the two lines do not exactly coincide. It is remarkable that they follow each other so closely. They differ because tree growth is limited not only by summer temperature, but also by drying winds, permafrost (i.e. permanently frozen ground) and poor soils. Curiously, extreme winter temperatures do not seem to inhibit tree growth; trees grow well in Subarctic Canada and Eurasia where winter temperatures are very low indeed. Several workers have tried to find other isotherms, or formulae linking summer temperature with humidity, wind, winter temperature and other factors, which match the tree line more closely than the 10°C. isotherm. Their work is more an effort to explain the tree line than to find a polar-subpolar boundary, and need not concern us here.

The tree line itself is not so simple as it may look. In atlases, it is a firm line between brown tundra and dark green forest, suggesting that trees march tidily to the tundra edge and stop there. In fact, it is often difficult to say within a hundred miles where the tree line lies. First you must define a tree, for species which form tall trees in the forest, produce stunted, shrub-like specimens on the forest edge. One sensible definition says that a tree must be tall enough to stand clear of winter snows. This

The simple life. Arctic poppy in a rock desert, Svalbard.

SIMPSON

allows field biologists to plot the tree line—either a single line for all species, or separate tree lines for each of the dozen-or-so species found at the northern forest edge. Even these tree lines are not straight-forward boundaries, for tongues and enclaves of forest penetrate far into tundra in sheltered valleys and hollows, and rising or swampy ground isolates islands of tundra within the forest edge.

Still, the tree line is a good, natural boundary. On a hillside it can be defined almost as sharply as on the atlas. At its fuzziest, on featureless rolling plain, it forms a zone several miles wide—some-times hundreds of miles wide—of stunted forest mixed with lichen and grassland, with scattered stands of true forest timber. This transition zone is important enough to be given a name of its own. "Taiga" is sometimes used for it, but has also been used confusingly for other communities of plants. "Forest-tundra" is descriptive, and I use it in this book. So the tree line and its natural expansion, the forest-tundra zone, together make our Arctic-Subarctic boundary. It is one which animals recognize; many are firmly committed to living on one side or other of the tree line. Arctic animals live on the tundra, and rarely, if ever, enter the forest. Subarctic animals are mainly forest creatures, but many move onto the tundra in summer. Only a few species are equally at home in both environments.

As page 24 shows, the Arctic so defined, includes the tip of northern Scandinavia, the northlands of European and Siberian U.S.S.R., the great penin-sula of eastern Siberia, the Pribilov Islands, Nunivak, St. Matthew and St. Lawrence Islands, the western and northern fringes of Alaska, and the northern edge of the Canadian mainland. The Canadian archipelago is entirely Arctic, with the northern half of Hudson Bay, all of Greenland, all but the southern coast of Iceland, the islands of the Arctic Ocean north of Eurasia, and isolated Jan Mayen Island.

From another of Köppen's climatological defini-tions we can include in the Subarctic zone all lands south of the Arctic in which mean temperatures do not exceed 10°C. for more than four summer months, and the mean temperature of the coldest month is not above freezing point. This gives a broad zone with a southern boundary running through the dark, boreal forest. The Subarctic is a zone of cool summers and cold winters. It need not be warmer than the Arctic in summer, though its mean monthly temperatures remain high (up to the modest limit of 10°C.—a cool spring day in London) for much longer. Nor need its winters be

Wind-sculptured trees of the forest-tundra, the limit of tree growth in northern Canada.

On mountain flanks the tree line is distinct. The tundra-forest boundary in the Canadian Rocky Mountains.

Bull moose, normally forest feeders, forage in scrub-tundra close to the tree line, Alaska.

warmer. Parts of the Subarctic are indeed much colder than the Arctic in winter. By far the coldest regions of the northern hemisphere lie not in the Arctic—which is warmed by the presence of the polar sea—but in Subarctic Canada and Siberia (page 20). The Subarctic zone includes the Aleutian and Komandorskiye Islands, central Alaska, most of Labrador and Newfoundland, and the southern coast of Iceland. Several major cities of Canada and Europe (Winnipeg, Montreal, Bergen, Oslo, Stockholm, Helsinki, Leningrad) lie just outside its southern border. Much of the Trans-Siberian Railway runs within it, linking many recently developed Soviet towns and cities.

Polar and subpolar seas

The climatic Arctic and Subarctic boundaries can be continued across the open sea, but oceanographers have ideas of their own about how the northern seas should be zoned. Their boundaries are based on physical properties of water masses. Arctic surface waters spread across the polar basin and through the Canadian archipelago. A long, narrow tongue of the same water stretches southward down the east coast of Greenland. This is recognizable by low salinity of less than thirty-four parts per thousand (due to dilution by northern rivers) and temperatures below −1°C. At its southern limit Arctic water meets and overrides warmer and more saline water, with salinity above thirty-five parts per thousand and temperatures above 0°C. On the Atlantic side this warmer water flows in from the North Atlantic Drift or Gulf Stream. A wide zone of mixed Arctic and Atlantic water stretches from the Kara Sea and Novaya Zemlya in the east to Baffin Bay, Labrador and Newfoundland in the west. On the Pacific side a weaker exchange spreads mixed Arctic and Pacific water eastward along the Alaskan shore, westward as far as Wrangel Island, and southwest to the southern shore of the Chukotskiy Peninsula of Siberia. These zones of mixing on either side of the polar basin are what oceanographers call Subarctic water. They are shown on page 20. It is worth noting that practically all of the oceanographer's Subarctic water lies within the 10°C. warmest-month

Autumn colour in forest and tundra. Barren Ground caribou in Mt McKinley National Park, Alaska.

Polar bear mother and half-grown cubs come ashore in summer to search for food. Coastal tundra of northern Canada.

BREUMMER

isotherm; if it were a land area, we would call it part of the Arctic. The southern limit of the marine Subarctic zone is the rather hazy boundary between the mixed water and the pure northern Atlantic and Pacific waters.

The small plants and animals of the plankton (page 54) which live in the ocean's surface waters need sunlight, mineral salts, and dissolved oxygen and carbon dioxide for their growth. On the success of the plankton depends the prosperity of nearly all other oceanic creatures, from shrimps and fishes to whales, from seabirds to bottom-living crabs, sponges and starfish. Pure Arctic surface waters contain plenty of dissolved gases but are poor in nutrient salts. Much of the Arctic Ocean is covered with pack ice (page 52), which reduces the amount of sunlight entering the water. So plant life is starved, and Arctic waters are only moderately productive. North Atlantic waters contain less dissolved gases but more nutrients, and are warm enough to disperse any ice which comes their way. Mixed Arctic and southern waters between them contain the best of everything; gaining the full benefit of continuous daylight in summer, they produce rich crops of phyto- and zooplankton (Chapter 3). These support large populations of pelagic or surface-feeding fishes, which in turn feed whales, seals and seabirds. So the boundary between pure Arctic and mixed Subarctic waters is biologically sound, and the one which is used in this book.

The northlands

The lands surrounding the north polar basin include ice caps, rugged coastal cliffs, alpine mountains, desolate marshes, estuaries, ice-bound beaches, and mile after mile of low-lying, featureless tundra. Only a fragment of them is permanently covered with ice; they are cold enough, but usually too dry to support sheets of land ice, and the few ice caps remaining are relics of the last glacial period.

Greenland is ice-capped; often said to be the world's largest island, it is actually three islands under a dome of ice nearly 3,000 metres thick. The cap is constantly being renewed by warm, snow-laden air from the north Atlantic, which deposits heavy snow also on Svalbard, Bear Island and Jan Mayen Island. Round its edge the cap is dammed back by high coastal mountains, leaving narrow ice-free plains exposed in the west, southwest and east. Northern Greenland is starved of snow, and has a wide belt of rolling, ice-free tundra. Heavy pack ice lies off the north and northeastern coasts. Baffin

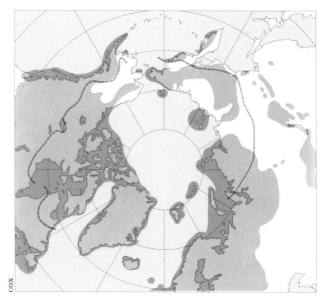

Shading shows the limits of ice sheets at the height of the Pleistocene glaciation. Red lines show the present limit of permafrost (permanently frozen soil) in the Arctic basin.

Bay is ice-covered in winter, but there is very little sea ice off the southwestern tip of Greenland. Warmed by the sea in winter, this is the mildest corner of the Arctic, with unusually rich tundra and forest-tundra vegetation. It has been settled and farmed at intervals for over a thousand years.

The remaining ice-free areas of Greenland have poorer vegetation, though enough to support local populations of musk-oxen, Arctic hares and other herbivores; caribou were once plentiful but have now almost disappeared. Greenland is well stocked with resident mammals and birds, and in summer becomes the breeding ground of migrant birds, which pour in from the south to take advantage of long hours of daylight and abundant plant and insect foods. The sea ice off Greenland supports large floating populations of breeding seals, polar bears and walruses, and open water feeds myriads of oceanic seabirds which nest on the high, coastal cliffs and offshore stacks. Whales have been hunted in Greenland waters since the Middle Ages, and extensive commercial fisheries have developed during the present century.

Iceland is a comparatively new island, formed by volcanic action within the last seventy million years and probably isolated from all other land masses since its first appearance. The volcanoes survived Pleistocene glaciations and are still active. New islets appear from time to time along the coasts, to the delight of local sea birds, whose colonies are overcrowded. About one eighth of Iceland is under permanent ice. Twenty thousand years ago it was

West Greenland ice cap, and a river formed by summer melting. This is one of the few really barren desert regions of the world.

Tundra lakes in August. Seemingly barren, these lakes teem with green algae, larval insects, fish and birds from May to September. Nordre Stromfjord, Greenland.

Volcanic Iceland. Warm springs produce clouds of steam and a patch of green vegetation, possibly with abundant insect life, close to stagnant glacier ice.

Fisherman's paradise. Like salmon, Arctic char leap the waterfalls of Iceland's shallow rivers to spawn upstream.

entirely ice-capped as Greenland is today. Since it lost its tremendous burden, the whole island has risen from the sea. Raised beaches, forelands and former submarine banks now stand high and dry along the coast, providing flat, fertile ground for human settlement. Shallow rivers with waterfalls cross the central plain, which is cold and barren.

Warm North Atlantic Drift water flows clockwise about Iceland bringing a mild climate to its southern shore. Good soils support a rich tundra and even forest vegetation. Timber felled by early settlers has not grown again, but fodder grasses flourish and cereals ripen in good years. Iceland has very few resident native mammals and birds, but like Greenland it supports a massive summer invasion of migrant geese, ducks and other birds; it is also a wintering area for migrants from colder regions of central and northern Europe, who prefer a maritime to a continental winter climate. Ice seldom appears in its coastal waters, and there is year-round commercial fishing on its rich offshore banks.

Northern Europe The North Atlantic Drift, which

brings warmth to Greenland and Iceland, flows on past Scandinavia to invade the polar basin. Its influence is strong enough to give Norway a Sub-arctic climate as far north as the 70th parallel. Eastern Canada, chilled by cold currents from the Labrador Sea, is Subarctic only to the 52nd parallel. In Europe only the northernmost tip of Norway and Finland, and the Barents Sea coast of Russia are Arctic. Scandinavia has a narrow strip of tundra coast with high forest-tundra inland. The Kola Peninsula is flatter and marshy, and the Arctic coast of eastern Russia has a wide belt of tundra and forest-tundra extending toward the foothills of the Ural Mountains. The sea remains ice-free for much of the year and the climate is mild. Spruce, Scots pine, birch and willow spread far to the north in sheltered valleys, and the open tundra is carpeted with grasses, shrubs and "reindeer moss", which is actually a lichen (page 100). This is good farmland, where crops grow quickly, and sheep, cattle and reindeer fatten on the warm summer pastures. There is a rich fauna of mammals, including lemmings, voles and other rodents, Arctic foxes,

SIMPSON

BREUMMER

The warming effects of the North Atlantic Drift. This lake near Kiruna, in the boreal forest of northern Sweden, is farther north (68 °N) than the tundra Barren Grounds of central Canada (right).

Reindeer feeding on waterlogged tundra of eastern Siberia. Permanently frozen subsoil (permafrost) keeps snow-melt water on the surface.

BOTTING

Early summer in the mountains of Arctic Norway. Melting snow reveals the remains of last year's vegetation.

BREUMMER

23

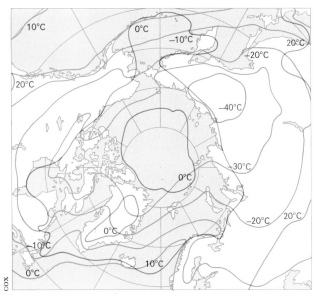

January (blue) and July (red) isotherms. The bulge of isotherms in the northeastern Atlantic is due to the Gulf Stream. Note that winter cold is most extreme in Canada and Siberia, not in the polar basin.

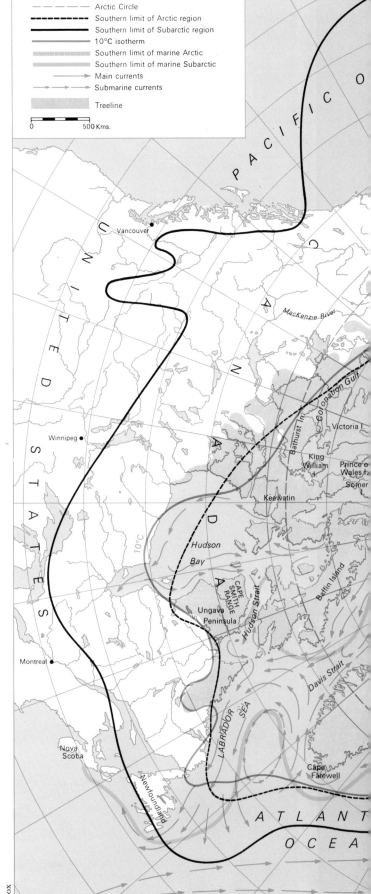

stoats, weasels and wolves. Many common European song birds spread northward into the tundra in summer, and the lakes and wetlands fill with migrant waders, geese and ducks.

The coastal tundra of northeastern Europe merges with the alpine tundra of the Urals, marked by a sharp southward dip in the Arctic-Subarctic boundary. To the north the fold of the Urals continues into low-lying Vaygach Island and the two mountainous islands of Novaya Zemlya. These islands are tundra-clad at low levels, with permafrost cementing the shallow soils. Surface melting in summer brings glacial torrents from the highlands, creating marshes and bog along the narrow coastal strips. The western flanks of the islands are ice-free for half the year, warmed by the easternmost tip of the North Atlantic Drift. The eastern flanks, facing the Kara Sea, are ice-bound even in summer. Reindeer are herded on the western slopes, and large populations of migrant geese and ducks breed there each summer. Polar bears and seals move in with the sea ice in winter.

The Siberian Arctic East of the Urals, the Arctic-Subarctic boundary turns northward, skirting the polar coast. Near the shores of the Laptev Sea, in latitude 73°N., are the world's northernmost trees

24

BERING SEA

KOMANDORSKIYE
ISLANDS

OKHOTSK
SEA

PRIBILOV
ISLANDS

Nunivak

Kamchatka Peninsula

St Matthew I.

St Lawrence I.

Kolyma River

DIOMEDE IS.

of Wales

Bering Strait

Chukotskiy
Peninsula

Cape
Lisburne

Wrangel I.

Lena River

NEW
SIBERIAN ISLANDS

ARCTIC OCEAN

LAPTEV
SEA

Severnaya
Zemlya

Taymyr Peninsula

180° 170°
160°   160°
140°   170°
120°   180°
100°   170°

North Pole

FRANZ
IOSEF
LAND

KARA
SEA

Northeast
Land

Novaya
Zemlya

West
Spitzbergen

Svalbard

URAL MOUNTAINS

BARENTS SEA

Vaygach I.

Bear I.

UNION OF SOVIET SOCIALIST REPUBLICS

Jan Mayen I.

Kola Peninsula

WHITE SEA

ICELAND

Kiruna

FINLAND

NORWAY

SWEDEN

Leningrad

Helsinki

Bergen

Oslo

Stockholm

Raised beaches, Bathurst Inlet, Canada. Relieved of its overburden of glacial ice, the land has risen in stages. Raised beaches show successive positions of the shoreline.

Barren Ground caribou crossing a broad river bed during their autumn migration from tundra to forest.

The indented coast of Labrador. Once heavily glaciated, this land has been flooded by rising sea level. Hard rocks yield scattered patches of poor soil supporting sparse vegetation and few grazing animals. A harsh polar environment in the latitude of Britain.

—stunted larch, spruce and alder, growing on dry, sandy patches in frozen marshlands. This is a cold coast, beyond the reach of Atlantic water and always ice-covered. Ice breakers keep the Kara Sea open for two or three months each summer. Further east the navigation season is shorter. The great Siberian rivers, though iced over for part of the year, carry heat northward from central Asia to soften the impact of winter on their estuary banks. Summers in the Siberian tundra are only slightly cooler than those of Arctic Europe; the isotherms run roughly parallel with the coast (page 24). Winters are very much colder. January mean temperatures are about −8°C. in northern Europe, −22°C. east of the Urals, and approach −40°C. in the cold heartlands of eastern Siberia.

Most of the coast is low-lying, with poor soils of glacial debris and shingle. The Taymyr Peninsula, half as big again as the British Isles, includes mountain ranges over 1,500 metres high, mountain lakes and broad marshy lowlands. Eastern Siberia is again mountainous, with only a narrow belt of coastal tundra. All of the Siberian Arctic is underlain with permafrost, in places over 300 metres deep. Only the uppermost metre thaws during the brief summer. Soils are dry and poor along the coast, richer inland toward the forest edge. Forest-tundra fills the sheltered river valleys, forming ecological through-ways several miles wide which bring forest animals deep into the coastal tundra in summer. Shrubs, grasses and mosses grow well inland, providing cover for many small mammals and birds and forage for reindeer herds. Coastal sea ice brings polar bears and seals inshore, bolstering the diet and economic resources of nomadic human populations.

Northern Alaska and Canada The northern and western tips of Alaska, the Keewatin, Ungava and northern Labrador peninsulas, and the northern archipelago of Canada are Arctic; so are the Pribilov, Nunivak, St. Matthew and St. Lawrence, and Diomede Islands of the Bering Sea. Southwestern Alaska is low-lying and marshy, with forest-tundra vegetation on low hills and islands of the Yukon River delta. Further north the coast is steeper, and the westernmost outliers of the Rocky Mountains reach the sea at Cape Prince of Wales and Cape Lisburne. Northern Alaska is a broad, rolling slope of dry tundra, descending in plateaux and shallow escarpments from the Brooks Range and fringed by beaches of marine sand and gravel. Though high peaks of the range retain fragments of Pleistocene ice, this area has been a dry desert for

BREUMMER

Weathered cliffs and scree slopes. Rapid weathering and extreme cold make it difficult for plants to become established.

BREUMMER

Kolunarn Island, off Baffin Island, Canadian archipelago. An area of poor grazing but rich sea where Eskimos live by hunting from the sea ice.

Brought to Svalbard by man, musk-oxen survive well on the bare tundra. Adventdelen, West Spitsbergen.

BREUMMER

many millions of years, with little history of glaciation. The coastal tundra is damped by fog and underlain by permafrost, which in summer yields a thin layer of ground water. Mosses, reindeer moss, grasses and herbs form a rich sward in summer, with a generous sprinkling of small flowering shrubs, dwarf willow and alder in sheltered corners.

The Canadian Arctic includes the Mackenzie River delta—a narrow strip of coastal plain flanking Amundsen and Coronation Gulfs, and the indented shore of Keewatin District, north and west of Hudson Bay. Bedrock is the hard granites, gneisses and ancient sediments of the Canadian Shield. During the Pleistocene glaciations, ice covered the whole area (page 46), depressing the land below sea level. Now the coast is rising and bears a thin covering of glacial drift, marine sands and gravels which make very poor soils. In Keewatin, these extend inland to Hudson Bay, forming a desolate region of lakes, bogs and poor tundra with scant vegetation. Across the bay, Ungava Peninsula is similar, with a narrow coastal plain flanking the Cape Smith Range. Northern Labrador, steep and rugged with an indented coast, is the southern tip of a chain of alpine mountains extending along the

Arctic meadow in spring. A core of permafrost three feet below the surface prevents melt water from draining away, flooding the tundra with streams, small tarns and shallow lakes. Mosquito larvae breed in tundra wetlands, producing food for many species of migrant birds. The migrants nest on dry ground between the lakes, protected from ground predators by the surrounding water.

eastern edge of the Shield. Ice-mantled and magnificent, these peaks form the backbone of Baffin and the southern bloc of Ellesmere Island at the head of Baffin Bay.

The Canadian tundra, like that of Siberia and Alaska, is underlain with permafrost. Freezing solid in winter when temperatures average −25°C. to −30°C., it becomes marshy in the summer thaw. A carpet of dwarf birch, willow and alder, reindeer mosses and grasses provides food for caribou, musk-oxen, Arctic hares, lemmings and ptarmigan. There are many predators, with both Indian and Eskimo hunters leading the field. A host of summer visitors from the south includes water birds and waders, which feed on the myriads of flying insects and their aquatic larvae.

The Arctic islands include a vast, complex archipelago north of Canada and a scattering of tiny islands and small archipelagos north of Eurasia. The Canadian archipelago has fourteen large and many small islands, totalling half a million square miles. Baffin Island, the largest, is a dissected chain of mountains rising above 2,000 metres. Still heavily glaciated, it has shed part of its Pleistocene ice load and is rising steadily from the sea. Banks, Victoria, King William, Prince of Wales and Somerset Islands lie west of Baffin and are generally low-lying, rolling and free of permanent ice. The Queen Elizabeth Islands form a northern group, including Melville, Devon, Prince Patrick and Ellesmere Islands. These are higher and several are ice-capped. Ellesmere, largest of the northern group, has peaks rising to 3,000 metres and more. Its northeastern coast is separated by only a narrow strait from the northwestern corner of Greenland, and its northernmost point is only a little over 400 miles from the geographical pole.

Seaways between the islands are icebound in winter and filled with drifting fast ice in summer. They include the long-sought Northwest Passage, a route which is still difficult for ships other than ice-breakers to negotiate. Permanent pack ice of the Arctic Ocean fringes the northern islands. Temperatures in winter average −25°C. to −30°C., rising to a little above freezing point in summer. The very dry climate supports only a sparse vegetation. Sheltered fiords of southern Baffin Island have the richest flora, and parts of southern Ellesmere Island, warmed by the open water of Baffin Bay, bear a continuous sward of mosses and grass. Small herds of caribou and musk-oxen roam the northern islands, and with Arctic hares, Arctic foxes and lemmings they are present all the year round. Geese,

Snow geese. Spring brings many species of migrant water birds to feed on the vegetation and insect larvae of the tundra wetlands.

Arctic ground squirrel in lush summer vegetation, Alaska. Ground squirrels burrow under the snow in winter, feeding at ground level on remnants of summer growth.

29

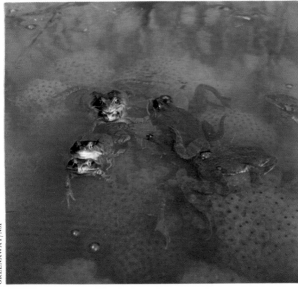

GREENAWAY/NSP

ducks and other migrants are plentiful in summer. The sea ice supports large populations of polar bears and seals, which provide the main food for Eskimo communities throughout the year.

Svalbard, a group of five large and many small islands including West Spitzbergen, Northeast Land and outlying Bear Island, was heavily glaciated in the Pleistocene era, and all its islands now wear remnant ice caps. Deep fiords and glacial

Frogs reach their northern limit just beyond the Arctic circle. Their geographical limit may be set by low water temperatures, which inhibit development of eggs and tadpoles.

Bull walruses sunning on an Arctic beach. Young walruses are highly sociable, gathering in clubs and feeding together on the sea floor. Later in life they scatter, tending to live in small family groups.

valleys dissect the larger islands. Soils are poor and barren, though there is enough lichen to support introduced musk-oxen and native reindeer. Lemmings, Arctic foxes and migrant birds make up the bulk of terrestrial animals. Huge sea bird colonies line the coastal cliffs, and seals and bears are plentiful on the surrounding sea ice. Svalbard is a Norwegian dependency with a population of miners, fishermen and fur-trappers. There are good deep-water fisheries close at hand.

Franz Josef Land, an archipelago of over a dozen, steep, ice-capped islands and many islets, is surrounded by heavy pack ice for most of the year. Beyond the warming influence of the North Atlantic Drift, these are truly Arctic islands with no permanent human occupants. Further east are the scattered islands and archipelagos of the Soviet Arctic, including Severnaya Zemlya, the New Siberian Islands, and lonely Wrangel Island on the 180th meridian. Nearly all are ice-capped and steep, colder than Franz Josef Land in winter, though still a little warmer than the continental coast to the south of them. Only a thin covering of barren soil has formed on the recently exposed rocks of these islands. Vegetation is restricted to the hardiest dwarf willows, grasses, herbs, mosses and lichens. There are very few land-based mammals or birds, apart from reindeer, lemmings, Arctic foxes and migrant water birds on some of the more favoured groups. Sea birds, seals and polar bears are more plentiful, and Wrangel Island has a permanent human population of hunters and fishermen.

Arctic plants and animals

In summer the Arctic is a pleasant warm place, climatically similar to temperate regions for two or three months each year. It is similar enough for many millions of birds to leave temperate latitudes, come flocking in to the Arctic and make themselves at home there. The discomforts of winter are usually overstated, though Arctic winters can be harsh, with consistently low temperatures, frozen soil and drifting snow. They are tediously long, tending to swallow spring and autumn and last from September to May. Yet, the Arctic has a perfectly habitable climate. Neither plants nor animals live under intolerable conditions, and there are plenty of both in all but the most desolate corners of the tundra. Greenland alone has 500 different kinds of ferns and flowering plants, and over 600 kinds of insects. Temperate or tropical regions of equivalent area

BREUMMER

Arctic terns experience two summers each year, flying the length of the world to the Antarctic coast between northern breeding seasons. They dip for small fish and crustacea in surface waters.

would have many more, but these figures help to show that many kinds of animals and plants can colonize an Arctic environment.

The simplicity of Arctic ecology is in fact based on the rather small number of species to be found in any one situation. This may partly be due to the harshness of the environment; it is certainly due to the *newness* of the environment. Many millions of square miles of Arctic terrain have emerged only recently from beneath the ice of the last glacial period. Some, indeed, are still emerging, as the land warms and the glaciers retreat. They have barely had time to thaw out, develop meagre soils, attract a sparse flora from neighbouring tundra, and offer their emptiness to animal invaders.

Moreover, the Arctic environment is new in another sense. Since the long-forgotten glaciations of the Carbo-Permian ice age (page 36), the world has simply not known such curious habitats for plants and animals as polar regions offer today. Except during the past two to four million years, there have been no ice-covered seas, no frozen soils at sea level, no solid-frozen lakes, nor grass plains snow-mantled for three of the four seasons. Nowhere have there been summers of Arctic coldness with a day lasting four months, followed by bitter winters of continuous night. These are entirely new habitats. In the history of the earth no bird, mammal or flowering plant has met anything like them before, for these forms had not evolved during the earlier ice ages. They are harsh habitats, for which plants and

OTT/COLEMAN

Red fox in summer fur, June. Like many Arctic mammals, red foxes wear a thin coat in summer, changing to denser fur in autumn.

OTT/COLEMAN

Red fox in winter fur, April, before moult.

animals can adapt—given time—but as yet they have not had time to mould many species to their own rather exacting requirements.

What are the origins of Arctic animals and plants? Like the inhabitants of any other new and emerging country, polar species are a mixed bag. Some are seasonal visitors which raid the Arctic in summer and leave it for warmer latitudes before the start of winter. Mostly birds, they are adapted for long-distance travel and have no special defences against extreme cold. Some are genuine natives—hardy remnants of the pre-glacial flora and fauna. Like native populations the world over they tend to be crowded out by recent immigrants; their environment has changed so rapidly that a distinguished ancestry gives them no special advantages in the struggle for space. Most polar species are simply immigrants—vigorous, determined representatives of temperate species, which, through competition and population pressures on their home grounds, have found profit in adapting to the Arctic. Many are recruits from alpine tundra, already adapted to cold, dry conditions with poor soils underfoot.

Established on the polar tundra, organisms have tended to find one part very much like the next, and to spread quickly about the polar basin. The mosses, lichens, shrubs, trees, insects, birds and mammals of Alaska have a strong family resemblance to, and are often indeed identical with, their counterparts in Canada, Greenland, Svalbard and Siberia. East-west winds, circulating sea ice, and the Beringian land bridge (page 44) have all

helped to give polar species a circumpolar distribution.

What additional adaptations have Arctic animals acquired for living in the extreme cold of the polar winter? Cold-blooded animals show very few adaptations which can immediately be recognized as tricks or devices for dealing specifically with the Arctic. Fishes and many invertebrates live successfully at temperatures very close to freezing point; some can concentrate their tissue solutions, allowing them to survive without freezing solid at temperatures well below 0°C. Many have longer life-cycles than their fellows in temperate regions; slowed by cold, their life-processes are drawn out over several years, where a temperate species would complete its cycle and die in a

Greenland collared lemmings avoid the extreme cold of winter by burrowing under the snow and living in a system of sheltered, insulated tunnels close to the relatively warm earth.

SIMPSON

single season. The higher cold-blooded vertebrates—amphibians and reptiles—tend to avoid the Arctic. Frogs reach the tundra in northern America but do not penetrate far, possibly because hatching and tadpole-development take too long in the cold water of Arctic lakes. Snakes and lizards seldom emerge from the forest; relying on direct sunshine to warm themselves and hatch their eggs, they are at too great a disadvantage during the long winters and cannot survive. Subarctic reptiles are viviparous, but even this device does not help them to penetrate far into the northlands.

Warm-blooded animals are conspicuously successful in the Arctic. Every bird and mammal of temperate latitudes must be well insulated against heat losses, and the additional insulation required for life in the far north does not make excessive demands upon them. Any warm-blooded animal with an assured food supply, cover and protection for its young, and a modicum of insulation can do well in the Arctic. Seals and whales, already adapted for life in waters much cooler than themselves, are completely at home in polar waters; seals have little difficulty in meeting the additional demands of life on the sea ice, so long as they can feed regularly to maintain their internal heat pro-duction. Birds also are pre-adapted for polar life. Small creatures, with an enormous surface area for their size, they face formidable problems of heat conservation even in temperate climates; having evolved some of the most efficient thermal insulation known to man, they wear it and go about their business with equanimity in the polar climate. Terrestrial mammals vary their insulation according to the season, as every trapper knows. Doubling and redoubling their fur in winter, hoarding or laying-on fat during the months of plenty, and living as quietly as possible during the long, lean season, keeps the larger mammals going through the polar year. The smaller ones—lemmings, voles and ground-squirrels especially—live close to the ground in a sheltered environment of their own, i.e. vegetation during the summer, and under the snow in winter. Though marginally fatter than similar species of warmer climates, they have no special adaptations for polar life—only the ability to take every advantage offered by their size and the nature of the habitat.

Huskies sleep out in the snow. Dense fur keeps in their body heat; snow settling on them does not melt, but forms an additional layer of insulation outside their coat. The long bushy tail curls over to protect the nose from frostbite.

The Changing Polar Climate

The ice cap, Svalbard.

Why are polar regions cold? There are two answers at least to this question. One tells us why the poles are colder than the rest of the world. Another, and a completely different one, tells us why they happen to be icy at present. To set these answers in perspective it is worth stressing that we are living at an unusual time. We are seeing the world during one of its rare ice ages. The last ice age—the Permo-Carboniferous ice age—ended 150 million years ago. There were possibly ten or a dozen ice ages before that during the preceding 3,000 million years of the earth's history, but, until our own began, there had been none since. Through most of geological time temperate conditions have extended around and across polar regions. As little as five million years ago dense forest and grassland extended north to the coast of an ice-free Arctic

Evergreen spruce forests south of the tree line provide shelter and food for moose, caribou and other herbivores in winter. Up to and during the Pleistocene period, dense forests spread north to the Arctic coast and covered many northern islands.

Ocean. Whatever the factors which keep polar regions cool, others have helped to make them icy, and to spread intense cold far into temperate latitudes of both hemispheres.

Solar radiation

Polar regions are by definition areas where the ends of the earth's axis intersect the earth's surface. They are always colder than the rest of the world because they receive less energy from the sun. Practically all of the earth's surface heat is derived from the sun's radiation, and the energy received is proportional to the angle which the surface presents to the sun's rays. Because the earth's axis is almost at a right angle to the sun's rays (page 37), the ends of the earth always receive their radiation tangentially. To an Eskimo, the sun never climbs high in the sky. Though it sometimes feels warm, it is seldom so warm as the tropical sun, which at times passes vertically overhead.

A column of sunlight one metre square, shining on a tropical land, takes the shortest possible path through the atmosphere and warms a square metre of earth intensively. In polar regions, a metre-square of sunlight takes a longer path through the atmosphere (which absorbs a great deal of it in passing) and spreads over far more than a square metre of ground. So it cannot warm the earth half so intensively. When the earth is cool, the air above it remains cool. However, just as late afternoon sunshine can feel warm in Britain, so the low polar sun warms polar lands locally in summer. South-facing rock slopes far north of the Arctic Circle sometimes reach temperatures of 40°C. to 45°C. at midday, so long as there is no wind. Patches of moss and lichen, dark and absorbent, are warmed intensively in the sun, producing tiny areas of semi-tropical conditions for the mites and insects living among them. If polar regions were permanently snow-free, their rocks, vegetation and ocean surfaces would absorb enough solar energy to provide warm temperate summers. They might absorb enough to keep the ground frost-free at sea level even in winter, giving reasonable year-round conditions for plants and animals. This we believe to be the normal condition for the earth's surface.

The poles are especially cold today however, because they are covered with ice. Permanent ice interferes with solar radiation in two ways. Firstly it is a mirror, reflecting back into space between 80% and 95% of the solar energy which falls upon it. In contrast dark rocks and calm blue seas

Pack ice has covered the north polar ocean intermittently through the past one to two million years. Early spring in pack ice, 81°N.

How the tilt of the earth gives us long polar days in summer, and short days without the sun in winter.

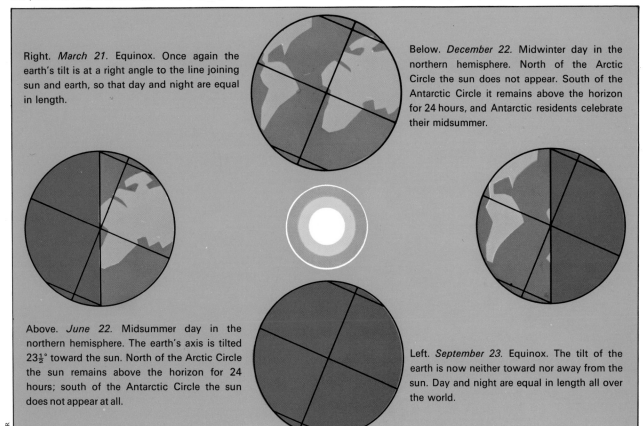

Right. *March 21*. Equinox. Once again the earth's tilt is at a right angle to the line joining sun and earth, so that day and night are equal in length.

Below. *December 22*. Midwinter day in the northern hemisphere. North of the Arctic Circle the sun does not appear. South of the Antarctic Circle it remains above the horizon for 24 hours, and Antarctic residents celebrate their midsummer.

Above. *June 22*. Midsummer day in the northern hemisphere. The earth's axis is tilted 23½° toward the sun. North of the Arctic Circle the sun remains above the horizon for 24 hours; south of the Antarctic Circle the sun does not appear at all.

Left. *September 23*. Equinox. The tilt of the earth is now neither toward nor away from the sun. Day and night are equal in length all over the world.

absorb between 80% and 90% of the radiation falling upon them. So the fifteen million square miles of polar ice and snow turn away between half and three-quarters of the solar energy which, in their absence, would warm the underlying ground or sea. Secondly, snow and ice use energy in vaporizing and melting. Air temperatures cannot rise far above freezing point while ice is present, because the solar energy is absorbed by the ice for melting, instead of being able to warm the environment. Where winter snowfall is heavy, summers tend to be short, for the air remains cold while the drifts of winter are dispersing. In drier regions with less winter snow, the ground warms quickly in spring.

Because of the ice caps, the polar regions each year lose more heat by radiation than they gain from the sun. Their energy budget is balanced by inflow of warm air and sea currents from lower latitudes. In the northern hemisphere the North Atlantic Drift carries heat from the tropical Atlantic direct to the polar basin, and the great Siberian rivers bring the warmth of central Asia to the Arctic shore. Atmospheric cyclones hurl masses of warm air into the atmosphere above the Arctic. But for these influences the Arctic would be very much colder than it is. The coldest part of the northern hemisphere is northeastern Siberia, where very little warmth can creep in from outside during the winter.

Formation of the ice caps

Why did the polar ice caps form? We do not know the full answer, but we do know that the present ice age did not begin without warning. A long spell of world cooling preceded it. Between the middle of the Mesozoic era, some 150 million years ago, and the start of the Pleistocene, three million years ago, the mean temperature of the earth's surface fell from about 20°C. to a little above 10°C. Cooling affected polar regions more than equatorial; the tropics may indeed have warmed slightly as temperate and polar regions cooled.

This change in world climates followed, and probably resulted from, massive changes in the distribution of land and water on the face of the earth. Up to 150 million years ago the continents were locked together in a single landmass (page 40). North America, Greenland and Eurasia together formed *Laurasia*, making up half of the super-continent. South America, Africa, India, Antarctica and Australia were united in *Gondwanaland*,

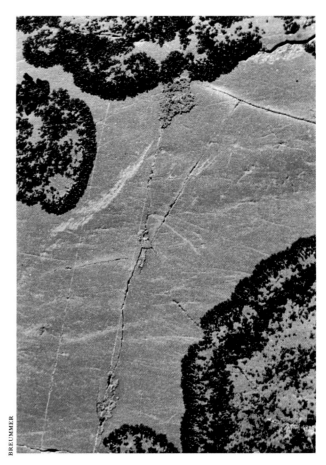

BREUMMER

Dark encrusting lichens absorb solar radiation, normally experiencing temperatures several degrees higher than air temperature while the sun is upon them.

White fur—the polar animal's personal greenhouse. Solar radiation penetrates deeply, warming the animal close to its skin.

BREUMMER

Mosses and lichens growing on an old spruce root. These communities absorb sunlight on warm, sunny days, providing subtropical conditions for insects living among them.

Polar seas remain cold because pack ice reflects up to 95% of the solar radiation falling upon it. Harp seals on a glistening mirror of one-year-old pack ice.

the other half. During the Mesozoic era Gondwanaland was rent by a system of fractures, and the separate continental blocs began to drift apart. Australia and the block which is now Greater Antarctica split away from the rest of Gondwanaland, forming the gap which widened to become the present Indian Ocean. About 120 million years ago South America began to split gradually from Africa, forming the Atlantic Ocean rift. Africa and India were pushed toward Laurasia, rotating slightly away from each other. Greenland and Canada began to drift apart at roughly the same time. The Atlantic rift spread northward, separating Europe from Greenland and opening the Arctic basin about eighty million years ago. The Americas swung westward, and Antarctica separated from Australia about sixty million years ago.

During this same period of disruption the continental blocs were shifting in relation to the earth's axis; the magnetic and geographic poles appeared to be wandering across the continents. About 400 million years ago the South Pole was in the position of the Sahara Desert, shifting successively during the late Palaeozoic and Mesozoic eras across West Africa, southern Africa, India, Australia,

Continental drift: four stages in the evolution of the world's continental masses.

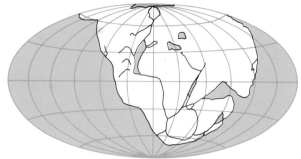

Pre-Mesozoic; up to 150 million years ago. Laurasia (north) and Gondwanaland joined in a single super-continental mass.

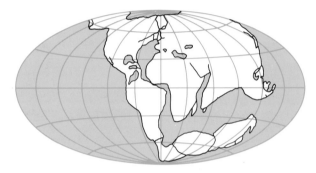

Mesozoic; about 120 million years ago. The Americas begin to swing westward; Antarctica and Australasia move south and east together.

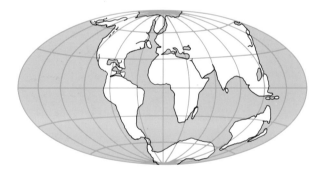

Early Tertiary; 60 million years ago. The southern Atlantic and Indian Oceans widen; Australasia and Antarctica separate and move into position.

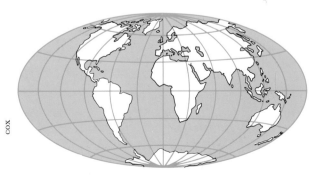

COX

Recent; The present distribution of the continents.

and eventually crossing a wide ocean to reach Antarctica. Its movements can be plotted by a study of the residual magnetism in ancient rocks. This reports the position of the magnetic poles, which are always closely associated with the geographical poles, in each geological period.

Wherever it moved over the continents, the pole was surrounded by a wide ring of glacial activity, which can now be detected from Carboniferous and Permian rocks of the Gondwanaland continents. The tropics during this long period formed a wide, shifting band across North America, Greenland and Europe. Temperate or tropical swamps gave rise to the coal measures which are now found as far north as Svalbard and Sub-arctic Eurasia.

Many climatologists believe that polar regions are cold only when they are surrounded by continents or land-locked seas, and that warm climates extend into high latitudes when oceans surround the poles. With the poles on land, warm oceanic currents cannot penetrate to sweep away intense cold. If the land is high, ice caps must inevitably form, and the polar regions become progressively colder. With the poles in mid-ocean, winter ice is dispersed and never allowed to become permanent. The tropics are cooled as the poles are warmed, and the earth as a whole becomes temperate. From 150 million to about seventy million years ago the poles are believed to have been located either on low ground or in the oceans, and the world seems to have lacked glacial activity altogether. Sixty to seventy million years ago the South Pole and Antarctic continent coincided in their travels, and the North Pole approached its present location in a land-locked sea. This is the time when cooling began in middle and high latitudes.

Both the southern continent and the lands surrounding the northern sea basin were low-lying at first, but the early Tertiary era of fifty to seventy million years ago was a period of mountain building. By the mid-Tertiary, extensive mountain ranges dotted Antarctica, and snowfields must certainly have formed among their peaks. Later mountain building on a world-wide scale produced new chains of highlands in Antarctica and gave the world its Pyrenees, European Alps and Himalayas. By the end of the Tertiary, mountain building is estimated to have lifted the world's continents from an average height of about 300 to over 800 metres. According to the climatologist C. E. P. Brooks, this elevation must have cooled the world by an average

BREUMMER

Melting snow and evaporating water use up solar energy, which would otherwise be available for warming the ground. Their presence keeps the Arctic cool in spring.

NORSK TELEGRAMBYRA

Coal mining in Svalbard. Coal measures in high northern and southern latitudes show that the polar regions of today were once the site of lush subtropical or warm temperate swamps.

The great Siberian rivers bring the warmth of central Asia to the Arctic shore. The Lena.

of 5°C., through increased cloud cover, evaporation, and the development of permanent, reflecting ice on the peaks of new mountain ranges.

So the world cooled through the Tertiary, slowly at first and then more rapidly. Still only five to six million years ago, the polar regions were temperate at sea level. Mixed deciduous and coniferous forests covered the plains of northern Eurasia and America, and tapirs, mastodons, browsing horses and sabre-toothed tigers roamed widely through woodland along the coast of an ice-free Arctic Ocean. What caused the final plunge into the ice age?

The period of cool, temperate climates immediately before the start of the ice age extended over several million years, and the glacial period itself has been one of fluctuating climates, rather than of steadily intensifying cold. It seems probable that, once the world had chilled to the threshold of a new ice age, another factor came into prominence, controlling both the onset of the glacial period and its succession of climatic changes. Several possible factors have been suggested. Among the most likely is the slow, secular fluctuation in the amount of solar energy which polar regions receive from year to year.

Milankovich cycles

This fluctuation is due to the combined effects of varying *obliquity* (or tilt) of the earth's axis, and the earth's *precession in orbit*. As the surface of the earth receives most of its warmth from sunlight, warming effects depend both on the angle between the sun's incident rays and the earth's surface, and also on the distance between sun and earth. Angle and distance vary over the centuries, in ways which cause cyclical fluctuations in the amount of radiation received in polar regions. The fluctuations could in turn cause cyclical changes in the climate of high latitudes.

Three fluctuations are involved, with cyclical periods of roughly 40,000, 92,000 and 22,000 years. Their combined effects have been calculated over a period of 600,000 years by the physicist M. Milankovich. This is roughly the period of the major glaciations in the northern hemisphere. There is a most convincing correlation between cycles of radiation maxima and minima, calculated by Milankovich, and cyclical fluctuations in climate during the same period, deduced by other workers from completely independent evidence. This is taken by many to indicate that the climatic cycles

BREUMMER

Glacier ice, the edge of a terrestrial ice sheet, Ellesmere Island. Up-to-date Eskimo with "tin sledge dog".

of the ice age were controlled by changes in radiation intensity.

Could such changes have started the ice age? The small variation in intensity of radiation implied by Milankovich's curves would not have started a climatic revolution in the equable conditions of a temperate world—the world of the late Mesozoic and early Tertiary. But in a world already cooled, with high land at one pole and partly-enclosed sea at the other, only a slight additional cooling would be needed to begin an ice age. The critical step or threshold is the formation of permanent ice caps from seasonal snow fields, a change which can be brought about by a very small decline in mean annual temperature. Once formed, ice fields are self sustaining. By reflecting solar energy away, they cool the world around them and spread. During the Pleistocene ice age, massive ice caps spread four times across North America and northern Eurasia. They could well have been started, and their subsequent spread and retreat could have been controlled, by the small fluctuations of climate induced when incident radiation varies.

At present we live in a warm interglacial period, not unlike that which preceded the start of the first major glaciation. Brooks has calculated that the extensive annual snowfields which now cover the ground each winter in Eurasia and North America are almost large enough to reach the threshold and become self-sustaining. They need only a slight climatic change—a succession of cool summers and early winters, or a slight lengthening of winter due to heavier snowfall (page 43)—to turn into vast, permanent ice sheets. This is the kind of climatic change which we believe can be invoked by variation in incident radiation. Once formed, the ice sheets would respond very slowly to further climatic fluctuations, perpetuating themselves, until a prolonged increase in radiant energy began to melt them.

A change of this magnitude could happen at any time, perhaps invoked by increasing atmospheric pollution or other human enterprise. If it did, the Subarctic would become Arctic. Many northern cities would be considered untenable, and vast areas of farmland would revert to forest and tundra. Human populations would shift southward, crowding an already teeming temperate world and leaning yet more heavily on overburdened tropical soils.

Slight changes could occur in the opposite direction. Longer and warmer summers are all that is required to melt the ice of the Arctic basin. In a minor way this has perhaps started already, for winters in Greenland and Svalbard have warmed by 5°C. to 10°C. in the span of two human generations. A continuation of this trend, helped along by man, could melt the polar sea ice within a further generation or two. This would result in milder winters over the northern hemisphere, longer and warmer summers, and good prospects for farming the richer areas of the tundra. It could also result in a catastrophic rise in sea level, due to the melting of the Greenland and Antarctic ice caps.

Summer snowfields in Alaska, with caribou keeping cool. Only a slight fall in annual mean temperature would be enough to convert these snow patches to permanent ice.

Animals and the ice ages

Seventy million years ago, at the start of the Tertiary era, Greenland was newly separated from Canada and Europe, the northern Atlantic Ocean was still widening as the Americas drifted westward, and Asia and North America were joined by a broad, low-lying plain, which geologists call Beringia. The Arctic plain spread north of its present coast, linking the polar islands with their mainlands. A great sea covered the southern plains of North America, and the northern plains were warm and damp. Beringia was heavily forested, with scattered, granitic islands, lakes and swamps. Cool, temperate forest, with willows, spruces, birch, pine and hazel, grew within a few hundred miles of the North Pole. Turtles and alligators, marsupials, insectivores, primitive carnivores and primates (lemurs) were widespread in northern forests of both the new and the old world.

Throughout the mid-Tertiary, temperate species of North America passed freely east and west across Beringia, and the plant and animal communities of America and Eurasia were almost identical. The Arctic Ocean formed a northern extension of the Atlantic Ocean, which continued to widen throughout the Tertiary. Completely different marine communities existed on the northern and southern shores of Beringia. Those of the north were Arctic, and clearly related to species of the North Atlantic. Those of the southern shore were Pacific, having evolved in the Pacific Ocean completely separately from Atlantic species. Even the seals of the two shores were different; the Arctic Ocean was populated by phocid or true seals (page 59) from the Atlantic, while Pacific shores were colonized by sea lions and walrus-like seals from the south.

Beringia persisted until ten to twenty million years ago, when the sea broke through briefly and Arctic and Pacific faunas mingled. A wide causeway again existed between Asia and North America during the early and middle Pliocene, five to ten million years ago. Fossil timber and pollen grains from this period show a change from warm to cold-temperate forests; spruce, hemlock, fir, pine and larch grew down to the sea, with birch, alder and only a scattering of hornbeam, elm and oak.

Right: Chart of the last 70 million years. Glacial conditions existed on earth about 150 million years ago, but were absent from the present north polar basin until two to three million years ago.

Far Right: the last 2½ million years. The four major glacial periods are shown roughly in proportion. The present "post-glacial" period of 10,000 years is slightly exaggerated.

44

TERTIARY AND QUATERNARY

Present ▶

PLEISTOCENE
Ice Age starts in northern hemisphere.

PLIOCENE
Northern forests lose deciduous trees. Bering Strait reopens, later closed. Cooling: first highland ice in northern hemisphere. Spread of cold-adapted mammals from Eurasia to America.

10

MIOCENE
Bering Strait appears briefly, later closed. Seals, sea lions, walruses evolved. Ice well established in Antarctic highlands. Deciduous woodlands still flourish on Arctic coast.

20

OLIGOCENE
Appearance of grasslands, and diversification of grazing, browsing and carnivorous mammals. First seals appear. World cooling: probable first appearance of Antarctic ice.

30

40

EOCENE
Spread and diversification of flowering plants. Warm climate on Arctic coast. First appearance of whale-like mammals.

50

60

PALEOCENE
Start of mammal diversification: flowering plants begin to dominate.

TYLER

70

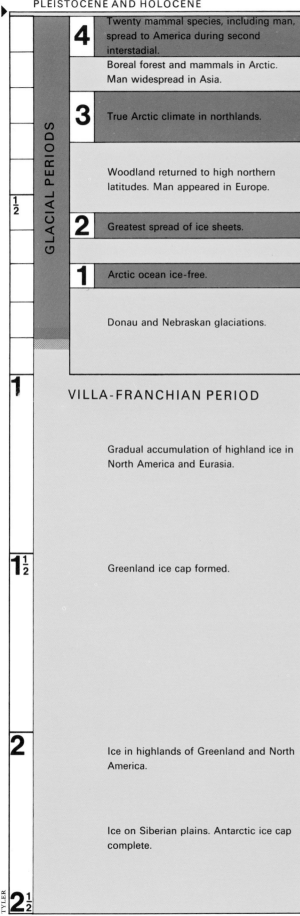

10,000 years

PLEISTOCENE AND HOLOCENE

GLACIAL PERIODS

4 Twenty mammal species, including man, spread to America during second interstadial.

Boreal forest and mammals in Arctic. Man widespread in Asia.

3 True Arctic climate in northlands.

Woodland returned to high northern latitudes. Man appeared in Europe.

2 Greatest spread of ice sheets.

1 Arctic ocean ice-free.

Donau and Nebraskan glaciations.

1

½

VILLA-FRANCHIAN PERIOD

Gradual accumulation of highland ice in North America and Eurasia.

1½

Greenland ice cap formed.

2

Ice in highlands of Greenland and North America.

Ice on Siberian plains. Antarctic ice cap complete.

TYLER **2½**

Beavers, primitive forest bears, chipmunks, tapirs, browsing horses and many small shrews and rodents roamed the forest floor throughout northern Eurasia and Canada. The forests were broken by steppe-like plains, which supported three-toed horses, rabbits and pikas (closely akin to rabbits), which were preyed upon by sabre-toothed tigers and other carnivores.

Between three and five million years ago the climate hardened. The northern forests lost many of their broad-leaved trees. Spruce, pine and larch remained, alternating with wide tracts of steppe-like grassland. Mammals continued to pass freely in either direction across Beringia. However, the late Pliocene and early Pleistocene saw the start of a new trend—a remarkable influx of Eurasian species into North America, with relatively few American species spreading westward into Asia. This one-way traffic was probably due to climatic differences between the two land masses. As the world cooled, the Asian heartlands would always have been slightly colder than North America, as they are today. Eurasian mammals would always have been the first to meet and adapt to the new conditions; having adapted, they would then have found themselves at an advantage when spreading eastward, in competition with their less-advanced counterparts in North America. The American fauna, always one move behind, would have found no advantage in spreading westward, where conditions were already colder than at home.

During this period the first deer crossed Beringia to the new world, accompanied for the first time by lynx, several new species of rodents (probably including the first lemmings), shrews, weasel-like carnivores, hyaenids, bears closely akin to modern grizzlies, and possibly mastodons. Later arrivals by the same route included mammoths, hares, wolverine, single-toed horses, shrub-oxen (a genus of browsing mammals now extinct), two new kinds of sabre-toothed tiger, and meadow mice, pine voles and squirrels. Many were animals of grassland and open country, for climatic changes were reducing the forest to steppe in the far north. Similar changes were affecting Europe; between the mid- and late Tertiary, coal-forming swamps and forests of northern Europe were gradually replaced by drier forest, later by coniferous forest-steppe. From the seas which covered south Britain and western Europe, coral reefs disappeared one and a half million years ago during the first really cold spell of the Pleistocene. The mammal fauna of Europe was a curious menagerie, including relics

45

of a warmer past (rhinoceros, hippopotamus, lion) and newly arrived species—mostly from Asia—which were better equipped to deal with the increasing cold; these included mastodons and elephants, zebra-like and horse-like grazing animals, fallow deer, *Megaceros* the giant elk, sheep, goats, bison and other ruminants, with sabre-toothed tigers, lynx and smaller dog-like carnivores.

The first permanent ice sheets of our current ice age formed over ten million years ago—perhaps as long as forty million years ago—among the high peaks of the Antarctic continent. Four million years ago they had joined, carved their way across the continental plains, and reached the sea. Two million years ago, calving icebergs were chilling the southern oceans and cold Antarctic water was creeping into the northern hemisphere over the ocean floors. Two to three million years ago the

first permanent ice sheets formed in Eurasia, and on high ground in Greenland, Iceland and North America. One and a half million years ago Greenland was completely ice-capped, and ice was accumulating on high ground all over the northern hemisphere. Then ice began to spread over the plains, slowly at first, advancing and withdrawing, from centres in Canada, Greenland, Scandinavia and the European Alps. The ice age had begun.

During the last million years the northern hemisphere has undergone four major periods of glaciation. They were first identified in Europe from studies of moraines and other glacial debris in valleys of the Alps, and were named the Günz, Mindel, Riss and Würm glacial periods, in order of occurrence. Ice sheets also advanced across northern Europe from Scandinavia, linking at times with those of the Alps and other centres. Three

Some earlier animals of the Arctic region. All were probably descended from subtropical species, which found their way north during warm periods. They adapted successfully when the climate cooled, and continued to flourish when extensive ice sheets covered the northlands. These and many other large mammals disappeared during the Pleistocene. Mammoths, mastodons, and woolly rhinoceros were hunted by man; other species may also have been killed or driven out by his activities.

Mastodon

Cervalces

Woolly rhinoceros

CASSELLI

46

Scandinavian advances have been identified—the Elster, Saale and Warthe-Weischel—correlating with the last three Alpine advances; there is no trace of an earlier Scandinavian glacial to match the Günz. Four glacials identified in America are called the Nebraskan (which started before the Günz), and the Kansan, Illinoian and Wisconsin, which correspond to the Mindel, Riss and Würm. Four have also been traced in Siberia, and given resounding names by Soviet geologists.

The Günz glaciation began six to eight hundred thousand years ago and lasted some fifty thousand years. There were two spells of intense cold (stadials) with a brief warm spell between. Günz ice sheets did not spread far from the Alps and northern Europe was only moderately chilled. In North America, Nebraskan ice covered the northern half of the continent, except for parts of western Canada and Alaska. The Arctic Ocean remained ice-free, and Arctic lands probably retained a forest-tundra or steppe vegetation. The first inter-glacial (i.e. warm period between glacial spells) lasted seventy thousand years. It restored forests and a rich mammal fauna to northern lands which the glaciers had denuded. Red deer, roe deer and moose (elk) were among the mammals which spread for the first time from Asia to Europe, with wolves following close at their heels. Primitive deer were also spreading across North America from Beringia, replacing the native prong-horn antelopes which retreated southward across the prairie as the cold advanced.

The second glacial period (Mindel-Elster-Kansan) also lasted fifty thousand years and in-cluded two stadials. In the second stadial ice sheets reached their greatest spread of the whole ice age.

Woolly mammoth

Giant elk

Sabretooth tiger

Britain was covered south to the Thames and Severn, and ice lay across Scandinavia, the Alps, the Low Countries, Germany and Poland. It reached south- and eastward over Russia and western Siberia, and a separate sheet covered the highlands in eastern Siberia and Kamchatka. Small ice caps spread over the Arctic islands of Siberia and Canada. In North America, ice from three main centres merged to form a single sheet, which spread south to New York, St. Louis, Topeka and the Missouri River, and west to the Canadian coast above Vancouver. Alaska and part of the Canadian coast remained ice-free, forming refuges for plants and animals. Greenland, Iceland and the Faroe Islands developed their own ice caps, which at times linked with those of America and Europe.

During this second glacial, Arctic lands were probably no colder than today, and may have been warmer. Sea ice was present in the polar basin, but the Arctic Ocean helped to warm the region as it does today, and the glacial period may not have lasted long enough to chill the northlands thoroughly. Europe gained its first reindeer, woolly rhinoceros, grazing elephants and other large mammals from Asia, and long-horned bison spread eastward from Asia to Beringia and North America. Man himself first appeared in Europe during the Mindel interstadial.

The interglacial following the Mindel lasted two hundred thousand years. Woodlands returned to high northern latitudes. Only a few, hardy remnants of the early Pleistocene mammal stocks returned with them, but the cold-adapted species of the late pre-glacial period were present in force. The third glaciation (Riss-Saale-Illinoian) lasted over one hundred thousand years. Though its ice did not extend as far south as before, this long glacial period brought the dry, cold, truly Arctic climate for the first time to the northlands. It ended nearly two hundred thousand years ago, giving way to an interglacial of seventy thousand years. In middle latitudes of the northern hemisphere this post-Riss interglacial was as warm as the preceding glacial had been cold, and woodlands once again spread far to the north. A rich fauna spread with it, including many of the mammals which were common in temperate forests and grasslands today. Pigs, deer, horses, bison and elephants grazed or browsed with rhinoceros, aurochs (ancestors of modern cattle), hippopotamus and rodents. Carnivores included lynx, lions, hyaenas, foxes, wolves, badgers and martens. Neanderthal Man was by this time well established in Europe and Asia, and already hunting the large mammals to destruction.

The final glaciation (Würm-Warthe-Weischel-Wisconsin) began just over one hundred thousand years ago and ended ten thousand years ago. There were three stadials, separated by one warm and one colder interstadial. During warm spells, the animals of central Europe included most of the species listed above. In cold spells these were replaced by Arctic foxes, wolves, hares, woolly rhinoceros and other cold-adapted species, whose home was in the north. During the second interstadial, summer temperatures in the northlands must have risen far above their present values. Northeastern Siberia, Beringia and Alaska were cool and dry, with little permanent ice present. The lowlands were grass-covered, with scattered woodland on higher ground. Some of the Arctic islands were forested, and the whole of the Arctic basin seems to have undergone a brief reversion to preglacial conditions.

At this time over twenty species of mammals crossed from Beringia to North America. They were mostly animals of the open plain and forest edge, and included practically all of the species which now characterize the Canadian Arctic—caribou, musk-oxen, Arctic hares, Arctic foxes, lemmings, collared lemmings and Arctic ground squirrels. Other species moved south to spread across the Subarctic and temperate zones, including moose, plains bison and other cattle, mountain goats, sheep, lynx, voles, meadow voles, and the first human invaders of North America. Woolly mammoths, two species of woodland musk-ox, two species of large deer and a steppe-antelope also crossed into the new continent and were swept southward in the final advance of the last stadial. The large mammals did not survive their association with man, whose hunting contributed to their extinction during the late Wisconsin and early post-glacial periods.

The fourth glacial reached its furthest spread just over twenty thousand years ago. It was followed by a slow, erratic warming which has continued spasmodically up to the present day. Between ten and twelve thousand years ago, temperatures in Europe had reached and even exceeded present values, and the ice sheets were rapidly retreating and disintegrating. A cold spell ten thousand years ago devastated new forests which were clothing the northlands, and Subarctic conditions returned to Europe. Eight thousand years ago this last dying spasm of the fourth glacial was long forgotten, and the north was again warming. Five to seven thousand years ago Europe passed through a brief warm phase, the "climatic optimum", with mean temperatures two or three degrees higher than today. Southern Britain developed warm, damp forests of oak, alder, elm and lime, and permanent ice may have disappeared altogether from the Arctic Ocean. Three thousand years ago a further cold spell developed. Mean temperatures fell slightly below present values, and pack ice returned to the polar basin.

Since then, temperatures of middle latitudes have fluctuated slightly on either side of a mean.

We have less information about Arctic climates, but they too are likely to be fluctuating. Folklore and historical records tell us that in medieval times southwest Greenland and Iceland supported farming communities which, during the late Middle Ages, fell on hard times because of climatic deterioration. Warming during the past century has helped to restore their prosperity. This is perhaps a slightly greater change than has occurred in Britain during the past two centuries. We are aware of oxen roasted on the frozen Thames two hundred years ago, Dickensian white Christmases in the mid-nineteenth century, warmer summers of the Edwardian period; these are oscillations similar to those which affected Greenland and Iceland, occurring at a comparable rate, though perhaps with smaller amplitude.

Tundra haymaking. Iceland, an isolated island with few native herbivores of its own, has been farmed for over a thousand years.

Young pack ice of a single winter off Ellesmere Island. Older pack ice is thickly snow-covered, corrugated, and usually several feet thick.

SIMPSON

Arctic Seas: the Chain of Life

The Arctic Ocean is one of the world's two cold oceans. With ice present its surface waters remain at about −1·8°C. to −1·9°C. Without ice surface temperatures are one to two degrees higher. Under summer sunshine shallow water in bays and fiords may rise to 3°C. or 4°C., warm enough to start seals and fishes moving seaward in search of a cooler environment. But the bulk of the polar ocean remains firmly within a degree or so of freezing point in winter and summer alike. Men have been known to die after five to ten minutes' immersion in water at these temperatures. Even the healthiest and strongest man can be chilled beyond recovery by half an hour's immersion in the Arctic Ocean. Seals and whales, though warm-blooded like man, are better insulated and adapted in many other ways to reducing heat losses (Chapter 6). So long as their fat lasts and food is abundant, they have no problem in keeping warm and active in icy waters.

Microscopic plants and invertebrate animals, which together make up most of the living material in the oceans, seem undismayed by the cold and indeed benefit from the stability of temperature in polar water masses. Their life chemistry is adapted to working at low body temperatures, in the same way as ours is adapted to working at 37°C., and most deep water species live in the certainty that, however low, their environmental temperature will not change by more than a fraction of a degree in their lifetime. As a result, many bottom-living species cannot stand temperature changes of any kind. Some polar fishes of deep water, which normally live within the range −1°C. to 0°C., die if their water rises to 2°C. Surface-living forms are usually more tolerant; to cope with seasonal and even daily shifts of temperature, they must remain efficient over a range of several degrees.

The freezing point of Arctic sea water is depressed below 0°C. by the salts it contains. Plant and animal cells usually contain less salts and would be expected to freeze at slightly higher temperatures. How do living organisms manage to remain unfrozen in polar seas? Experiments with Arctic and Antarctic fishes have shown that their tissue fluids are usually supercooled; their blood contains an anti-freeze compound which normally prevents ice from forming within the cells, but cannot cope if the fish comes into direct contact with ice crystals in the water. This may help to explain why so few fishes are able to live at the surface in Arctic waters, where ice crystals are often present. Invertebrates tend to have higher concentrations

of salt in their blood, and a correspondingly lower freezing point.

The central core of the Arctic Ocean is frozen permanently, covered with heavy *polar pack ice* which circulates year after year about the polar basin. It reaches a thickness of three to four metres. From its edge grows an annual ice sheet about one metre thick, which breaks constantly to form a complex of interlocking floes. This is *annual pack ice*. *Fast ice* grows each year from the land, sometimes spreading sixty to seventy miles out to sea. "Fast", or attached to the coast in winter, it breaks up in spring to join the annual pack and sweeps southward to melt in the northern Atlantic and Pacific Oceans. Polar pack ice has few animals, though bears, foxes and seals are sometimes reported close to the pole. Fast and annual pack ice are a floating platform for bears and seals, taking them into waters where food is plentiful and making them independent of land. The ice cover is never complete. Circulation and movement, currents and tides, cause leads and pools of open water—called *polynyas*—to form, especially over shallows and near headlands. These allow the sea to warm the local atmosphere and land in winter, and give seals and sea birds a chance to begin breeding in high latitudes before the main sheets of fast ice have broken away early in spring.

The plants and animals of the Arctic Ocean form interacting communities (page 54). First in importance is the community of the *plankton*—the small-to-microscopic organisms which live in the surface layers in summer. On these all other life in the sea depends. Plants of the plankton are collectively called *phytoplankton*; they float passively

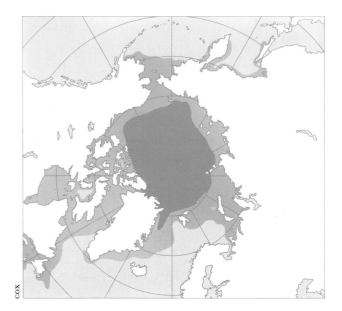

Leads and holes in the sea ice, formed by tension and tidal movements, allow seals to penetrate far inshore under the annual fast ice. Harp seal coming up for breath.

Below, left:
The southern limits of permanent pack ice (dark blue) and seasonal pack ice (medium blue) in the Arctic basin.

Floating home for seals and walruses. Remnants of summer pack ice in Hudson Bay, eastern Canada.

in the surface currents, using sunlight as a source of energy for their growth and multiplication. *Zooplankton* includes the small animals of the plankton, which feed on the plant cells and on each other. They too drift with the currents, but many are capable of small swimming movements which keep them at the surface.

Nekton is the community of free-swimming animals, including fishes, seals, whales and squid, which live both in surface waters and at depths. Those which prey on the plankton are said to be *pelagic* or surface-living. A few—very few in Arctic waters—live in mid-water, between the surface and the bottom and are called *bathypelagic* communities. Many fishes are *benthic* or *demersal*, living close to the sea bed and feeding on the benthic communities of plants and animals which cover it. In very shallow water approaching the shore, demersal communities become *sublittoral*, and eventually *littoral* or *intertidal*. Temperate regions have rich sublittoral and intertidal zones, with an extraordinary variety of plant and animal forms. In polar regions these zones are relatively bare. Few plants or animals can withstand the repeated freezing and thawing, buffeting by ice, and other hazards to which a polar shore is exposed.

Plankton

In polar seas, as in every other environment, living creatures gain their energy directly or indirectly from sunlight. This is not specifically the warming effect referred to earlier (page 36); the energy used by plants and animals is trapped far more effectively by the green pigment, chlorophyll, of living plants.

Phytoplankton is made up almost entirely of single-celled plants—mostly flagellates or diatoms—which in summer form a diffuse suspension in the top few metres of surface waters. Like all other green plants, these cells absorb mineral salts and carbon dioxide from their environment, and use the energy of sunlight to build up (photosynthesize) food materials. Oxygen, which is a by-product of photosynthesis, reliberates the energy chemically in a form which the cells can use for growth, movement and reproduction. The largest cells of the phytoplankton are about one millimetre across, as big as a pin head. The smallest are one thousand times smaller, invisible to man, though their presence colours the water. Rich concentrations of phytoplankton often appear as greenish or reddish masses in surface waters. Some are dense enough for

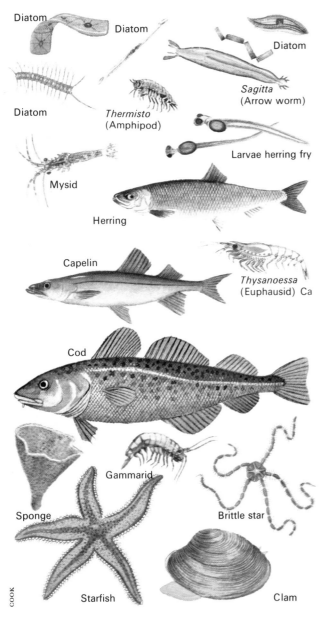

COOK

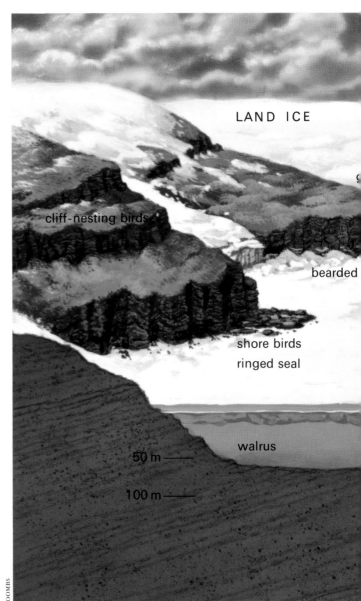

COOMBS

Food chains in the sea.

Diatoms – single-celled plants in the upper layers of the ocean – are especially numerous in late spring and summer. In sunlight, they synthesize the food materials on which all other organisms in the sea depend. Diatoms are eaten by both adult and larval *crustacea* in surface waters, whose mouthparts have tiny combs of bristles to filter the microscopic diatom cells from the water. *Larval fishes* also feed on diatoms. *Arrow worms*, many adult crustaceans and several species of fishes (e.g. *herring, capelin*) feed on the surface-living crustacea, in huge shoals which concentrate in the plankton swarms. They are joined by flocks of sea birds, and by plankton-eating seals and whales. The surface-feeding fishes are also taken by whales, seals and some of the larger birds, and man as top predator takes his pick of all fishes, mammals and birds. The constant activity in surface waters produces a rain of debris – mostly dead and dying animals and part-digested plant cells, which become the food of bottom-living fishes and crustaceans (e.g. *cod, gammarid shrimp*), and of *sponges, starfish, brittle-stars, molluscs,* and other creatures on the sea bed.

The ecology of Arctic shorelines and seas. Source of all energy is the sunlight trapped by minute floating plants of the phytoplankton. These flourish in leads and cracks between ice floes and in open sea, and are eaten by planktonic animals (zooplankton) which in turn become the food of birds, fishes, seals and whalebone whales. Seals and toothed whales also eat fishes, both in surface waters and in deeper water. Sea birds breed ashore, spreading valuable nitrogenous fertilizers on the vegetation near their nesting grounds. Seals breed on the ice or on shore, whichever is closest to their particular food supply. Whales give birth in the water, usually swimming to warmer latitudes and returning with their half-grown young.

FAST ICE   PACK ICE

gulls   skuas   sea birds

harp seal   polar bear

grey seal   killer whales   hooded seal

plankton   dolphins

pelagic fish   whalebone whales

clams

demersal fish

clams   sponges   starfish   worms   demersal fish

the waste products of the cells to form poisonous concentrations, which kill fishes and other creatures engulfed in the shoals.

Arctic waters are not especially rich in nutrients (page 20), and the concentration of phytoplankton is usually kept well in hand by browsing animals. These are mostly crustaceans (page 54), which use combs, nets or filters to sweep the plant cells into their jaws, or larval fishes which suck the sea water and plant cells in. Much as sheep and rabbits keep down a field of growing grass, so the herbivores of the zooplankton keep down the phytoplankton. Starting in May or June, the increasing hours of spring and summer sunshine stimulate the plant cells into rapid growth and reproduction. The grazing animals profit from the increasing turnover, growing and multiplying in their turn to

form huge shoals several hundreds of metres across.

Plankton at its richest is a thin, salty, vegetable consommé, with a scattering of minute fish and shrimp particles which provide added protein. Served in a restaurant it would be considered poor fare, with one small shrimp to every five or ten platefuls. But the fishes and marine mammals which dine off the plankton reject the stock and concentrate on the protein. They carry their own strainers (the baleen of whales, the interlocking teeth of plankton-feeding seals), and by filtering off the protein are able to make a solid meal and an excellent living while the soup lasts. In the high Arctic it does not last long. The nitrates, phosphates and other nutrients available in surface waters at the start of the season are used up within about five weeks, and plant growth ceases. There is very little mixing with sub-surface waters to bring new supplies of nutrients into circulation. So the Arctic plankton season is a short one. In Subarctic waters there is more mixing, and the plankton remains active and rich until August or September. This gives the marine birds and mammals a longer season in which to rear their young.

Though the plankton feeders adjust their breeding seasons to match the annual cycle of plankton activity, they are pressed for time. As the sun weakens in autumn the plankton begins to descend into the deep water. Fish follow it, but the birds and marine mammals which depend on plankton must move south for the winter. Late starters and laggards lose their young, for there are no second chances in the Arctic year.

The larval and juvenile animals of the zooplankton are mostly the young stages of adults which live sedentary lives on the sea bed. Joining the plankton allows the young to travel further afield and, more specifically, to disperse from the adult concentrations and to live on foods for which they are not competing with their parents. Other creatures of the zooplankton spend their whole lives in surface waters, feeding on different sizes of plant or animal prey as they grow, and finally liberating their own eggs or young into the plankton. Very few planktonic animals drift helplessly. Nearly all contain oil globules or gas bubbles, and are covered with spines or equipped with swimming organs, which help to keep them up in the water and control both their depth and the direction of their movement. By adjusting depth and keeping to the right currents, they travel in prescribed circuits which give them the best chances to survive and prosper.

Plankton shoals are surprisingly active, noisy communities. Visible at the surface, they attract flocks of sea birds which eat either the zooplankton itself or the fish which are feeding on it. Through a submarine microphone one can listen in to the sounds of a plankton shoal—a constant rustling roar, like a football crowd half a mile away—made up of the clatter and crackling of innumerable jointed limbs and tails, and the munching of myriad jaws. Commercial fishermen and whalers watch the birds, and use both echo-sounders and hydrophones to monitor the movements of the shoals, for plankton, fishes and whales move together. Seals, whales and other natural predators have both echo-sounding and sonar techniques of their own, no less sophisticated than man's, and probably as efficient in searching out their prey.

The sea bed fauna

The intense activity of the plankton and its predators in the upper layers of ocean produces a steady seasonal rain of debris. This includes dead plant and animal matter, part-digested food, faeces of all kinds, shells and skeletons, which fall to the sea floor. It becomes the food of demersal or benthic animals, which live in the dim world where sunlight may never penetrate and no other source of energy is available. Some of the material is broken down to its constituent minerals by bacteria, and dissolved in sea water. Much is ingested by larger animals—nemertine worms which ingest it whole, polychaet worms which spread delicate nets to pick up small particles, clams which suck water in and filter off the organic debris, and sponges which, by similar methods, process hundreds of gallons of silt-laden water per day. The insoluble, indigestible residue is left to form a carpet on the ocean floor. Polar plankton produces a distinctive pattern of carpet, called diatom ooze, which is made up almost entirely of the hard, siliceous shells of diatoms—the plant cells which dominate the plankton. Decomposition and decay on the ocean floor liberate the nutrient salts essential for life, and allow them to circulate and promote new activity in other parts of the ocean.

Seaweeds depend on sunlight for their living and cannot grow below depths of about 100 metres. Though sunlight penetrates farther (it can just be detected at about 1,000 metres in clear water), too much is absorbed in the surface layers to promote photosynthesis between eighty and 100 metres, and seaweeds which grow as deep as this are unlikely to flourish. Animals accept no such limitation; they have been found at the greatest depths of every ocean, and in polar seas form a thin, scattered community on and within the mud of the sea bed. They form the food of predatory fishes, which browse and nibble constantly over the surface and churn up the mud to expose its inhabitants. Several species of seals are bottom feeders too, notably the walrus, which ploughs the sea bed with its tusks in search of clams and other bivalve molluscs.

Fishes

Arctic nekton includes several species of fishes, nearly all of them demersal and usually spread between the Arctic and Subarctic marine zones. They include bullheads or sculpins (Family Cot-

BREUMMER

Bering wolf fish, a typical Arctic bottom-feeding fish. Crushing teeth and powerful jaws make mincemeat of molluscs, crabs, and other hard-shelled foods.

Arctic char, caught at the river mouth, gutted and drying in the sun. This is an important addition to the Eskimo's winter diet. Bathurst Inlet, Canada. See also p. 22.

BREUMMER

tidae), eel-pouts (Zoarcidae), sea-snails (Liparidae), several species of cod (Gadidae) and others, which browse on the benthic carpet of the polar basin. Polar fishes tend to produce large, yolky eggs, which they lay in shallow water. The growing larvae move up into the plankton for the summer, then sink to live the rest of their lives in deeper water.

One pelagic or surface-living species of high latitudes is the Arctic char, a handsome salmon-like fish, which divides its life between fresh water and the sea. Hatched in lakes and large ponds on the tundra, the young fish live in fresh water for several seasons. They descend to the sea in late spring or early summer of their fourth or fifth year, there to grow fat on plankton and concentrations of young fish— especially capelin (see below). After a summer spent in shallow coastal waters they ascend the rivers again and are ready to mate and spawn on reaching their lake breeding grounds. Fat char entering the rivers are caught in thousands by Eskimos, who appreciate the richness and delicate flavour of their reddish flesh.

Subarctic waters are richer in plankton and have a longer productive season. There is a greater variety and concentration of fishes, again with emphasis on bottom-living forms. Several deep-water species move into coastal or surface waters in spring to fatten and lay their eggs. Subarctic, demersal fishes include polar cod and halibut, which are smaller than their commercially important kin of the northern Atlantic and Pacific oceans, but occupy a similar niche on the sea bed. Fishes seem to be especially sensitive to temperature changes, and the steady warming of recent years has extended the northward range of Atlantic species into Subarctic waters. Atlantic cod, halibut, coalfish, ling, haddock and saith are now caught far to the north of their nineteenth-century haunts, bringing prosperity to Iceland, Greenland and other Arctic settlements. Greenland sharks— sluggish, bottom-living creatures up to five metres long, which live in deep water—have been caught locally for many years by Eskimos. Now they too form the basis of an industry, yielding abrasive skin and lubricating oil.

Capelin are small, herring-like fish which feed on plankton and form huge surface shoals in the shallow waters of Labrador, Greenland, Iceland, northern Europe and the Bering Sea. They winter unobtrusively in deep water, but come to the surface in May, June and July to feed and spawn in sheltered bays and fiords. Their immense shoals attract Arctic char, cod, seals, sea birds and even whales, which follow them into the shallows feeding voraciously. Occasionally, masses of capelin are driven by wind and predators into the inter-tidal zone, where Eskimos and other local folk scoop them out of the water by the thousand. After July the capelin disappear, leaving clusters of yellow eggs among the seaweed.

Whales

The whales of the world fall into two major groups, both well represented in the Arctic. Whalebone whales (Mysticetes—which means moustached whales) are generally large, ranging in length from ten to thirty metres. They are especially adapted for feeding on plankton. Lacking teeth, they possess instead a filter of whalebone or baleen, which allows them to strain zooplankton from the sea. Toothed whales (Odontocetes), which include dolphins and porpoises, are generally smaller. Sperm whales are the largest Odontocetes, bulls reaching lengths of eighteen to twenty metres. Killer whales are next in size, bulls averaging ten metres; most other toothed whales and dolphins are between two and six metres long. They feed mostly on fish and squid, some (especially killer whales) taking larger prey.

Whales are air-breathing mammals, lively, intelligent and extraordinarily well adapted for their aquatic way of life. Originally quadrupedal like other mammals, they have lost their hind limbs altogether. The forelimbs are used for hydro-planing and steering, and the horizontal tail flukes form a powerful rotary propeller. The skin is rubbery, and hairless except for a few sensory whiskers on the face. Whalebone whales and the smaller toothed whales conduct most of their business at the surface of the water; they feed, court, mate and suckle their young at the surface, and sleep with their blowhole just above the water. All the large whales are able to dive deeply and stay below for several minutes—even for longer than an hour. Bottlenosed and sperm whales probably dive more often than other species in search of their food. Odontocetes have a well-developed system of echo-location, by which they find and identify distant objects in dimly-lit or murky water. Their squeaks, groans and high-frequency clicks sound continuously, and they signal to each other by whistling, quacking and mooing. Whalebone whales, busily gulping plankton, seem to have less to say to each other.

Blue whales, which are actually a pleasing shade of grey, often reach lengths of twenty-five metres and may exceed thirty metres; large ones weigh more than one hundred tons. They are by far the largest animals which have ever existed, and among the most graceful. Fin, sei and minke or piked whales are smaller members of the rorqual group; grey, right and humpback whales are closely related Mysticetes, which amble more slowly through the sea but feed in the same way. All have enormous mouths, up to one third of the total length of the animal, containing a sieve of baleen plates suspended from the roof. The lower end of the plates is frayed, so that the roof of the whale's mouth seems to be lined with dense coco-nut matting. Swimming in pods or family groups, whalebone whales plough rapidly through the plankton shoals taking in huge gulps of sea water, which they press through the matting, leaving the zooplankton behind. A muscular tongue and ever-open throat take care of the residue. Taking in air through their blowhole or nostril, they breathe out a warm jet of mixed gases which turn to visible vapour in the cold air.

Whales are warm-blooded (page 33) and main-tain body temperature partly through sheer bulk, partly by the layer of blubber which underlies their skin. Blubber contains a clear oil which, until the development of mineral oils, was a useful fuel and lubricant. We still need it—or take it when we can—to make soap, margarine and cosmetics. What man needs no whale may keep, and whales have been hunted for their blubber (and other commodities: Chapter 6) for over three centuries. Right whales and humpback whales were hunted almost to extinction, the former in their Arctic feeding ground among the loose summer pack ice, the latter further south; hunting for these species finally became unprofitable late in the nineteenth century. The hunt for rorquals began in 1864 with the invention of the explosive harpoon and the development of fast, powered, whale-catching ships. Within a century blue whales had been almost completely eliminated from both ends of the earth, and fin, sei and the smaller species are rapidly disappearing as hunting pressure is con-centrated upon them.

Odontocetes, too, migrate towards the polar regions each summer. Solitary sperm whale bulls are occasionally seen in the Bering Sea and off the coasts of Labrador and Iceland. Their narrow lower jaw, with its row of peg teeth, is well adapted for taking slippery squid (cuttlefish). Sperm whales

BREUMMER

Greenland shark — one of the very few kinds of shark living in polar waters — is hunted by Eskimos for its oil and tough abrasive skin. It has been hunted commercially for several years.

have also been known to take fish and even sharks. Killer whales are the largest of the dolphin family. They have a distinctive pattern of clean white and black patches, a tall dorsal fin, which stands above the water like a sail, and a row of twenty business-like teeth in each jaw. Killers hunt in packs of twenty to forty individuals and have acquired an extraordinary reputation for ferocity. Packs of killers are said to attack rorquals and other large whales, roll seals off ice floes, and plough through concentrations of seals and dolphins leaving a trail of destruction. One killer is reported to have contained the remains of twenty-seven porpoises and seals in its stomach. They are widespread across the world, and fairly frequent visitors to Arctic and Subarctic waters in summer.

Narwhals, and belugas or white whales, are restricted to the Arctic and Subarctic. They are large dolphins, similar to each other in shape and reaching a greatest length of about five metres. Male narwhals are distinguished by their ivory tusk, a single tooth (usually a left upper incisor) which forms a straight, spiral rod about half as long as the animal. Nobody knows what the tusk is for. It is usually the only tooth which the narwhal possesses, and he does not seem to use it in attack,

defence or feeding. The tip is usually worn, suggesting that it may be used to stir food from the mud of the sea bottom. Narwhals and belugas feed mainly on fish and squid, and are small enough to be hunted from kayaks by Eskimos. They are a source of skin, sinew, ivory, oil, meat and bone.

Seals

Of the thirty-two species of seals alive today, ten live within our Arctic area or close enough to be frequent visitors. Like whales, seals are carnivorous mammals adapted for life in water. Unlike whales they are not fully aquatic; they produce their young ashore or on sea ice, and spend long spells out of the water between hunting trips. There are three families—walruses (Odobenidae), fur seals and sea lions (Otariidae), and the true or phocid seals (Phocidae), which are most completely adapted for life in water. Fossil and other evidence indicates that the walruses, fur seals and sea lions evolved during the mid-Tertiary from bear-like ancestors, probably along the Pacific coast of

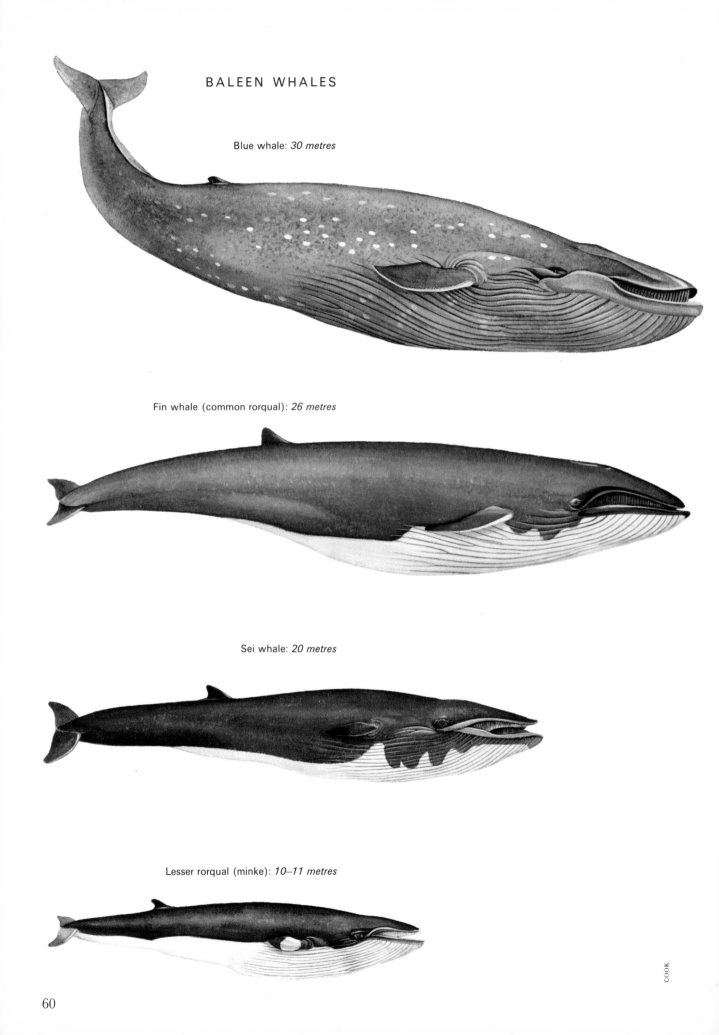

BALEEN WHALES

Blue whale: *30 metres*

Fin whale (common rorqual): *26 metres*

Sei whale: *20 metres*

Lesser rorqual (minke): *10–11 metres*

COOK

60

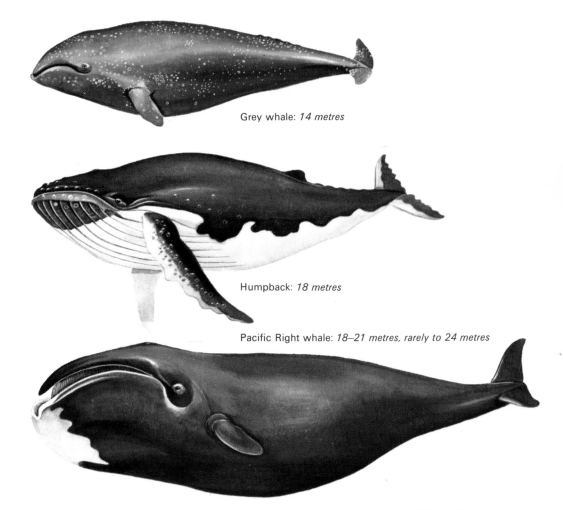

Grey whale: *14 metres*

Humpback: *18 metres*

Pacific Right whale: *18–21 metres, rarely to 24 metres*

The whalebone whales (left and above) are generally larger and live close to the surface. The huge head – one third of total body length – contains the sheets of baleen through which planktonic animals are filtered from sea water. Toothed whales, mostly smaller, feed on seals, fish and squid. Many of the whalebone whales have been hunted commercially to the verge of extinction. Among toothed whales, only sperm whales are large enough to be worth hunting commercially, but the smaller species (including dolphins) are taken locally by inshore hunters.

TOOTHED WHALES

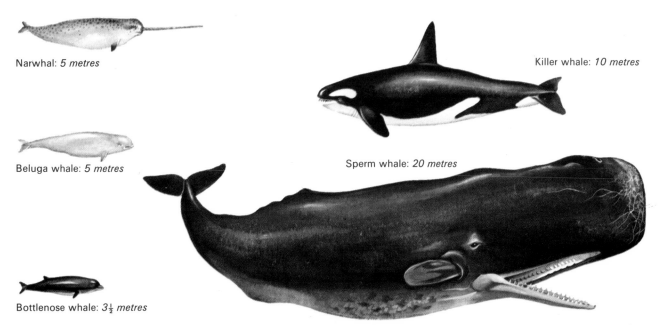

Narwhal: *5 metres*

Beluga whale: *5 metres*

Bottlenose whale: *3½ metres*

Killer whale: *10 metres*

Sperm whale: *20 metres*

North America. In spite of their paddle-like limbs, they still have a strong resemblance to bears or dogs. Fur seals especially have many dog-like characteristics; for yapping there is little to choose between a fur seal pup and a Pomeranian. Phocid seals more probably descended from otter-like animals of central Asia, which spread to the Atlantic and Arctic coasts during the Miocene. Though clumsy, seals can be surprisingly agile on land. Sea lions and fur seals run and scramble over rocks with their bent flipper-like feet, and phocids wriggle like monstrous but effective caterpillars over sea ice. In water they move exquisitely, rolling, turning, diving and leaping with effortless grace.

Walruses are entirely Arctic, while fur seals, sea lions and phocids have spread across the world from northern to high southern latitudes. Phocids have spread furthest; they are the typical seals of pack ice and fast ice in both hemispheres, and are well represented in the tropics too. Fur seals and sea lions are mainly subpolar, overlapping in range with phocids but avoiding the coldest water and keeping well clear of permanent ice.

Walruses are the heavy, lugubrious seals of the North Polar basin and Bering Sea. Large male walruses measure over four metres from moustache to tail and weigh a ton and a half. Females, more delicately built, weigh up to three-quarters of a ton. They live close to the coast, feeding in shallow water and hauling out onto the shore in summer and the sea ice in winter. Walruses of the northern Pacific, Alaska and eastern Siberia have a heavier skull and longer tusks than those of eastern Canada, Greenland and northern Europe. The two stocks seem to be geographically isolated, and are usually regarded as separate subspecies.

With their bloodshot eyes, heavy jowl, bristling moustaches and permanent air of grievance, walruses caricature a particular brand of human pomposity; it is difficult to take them seriously. However, their funny faces are remarkably functional. The massive head is used in breaking through newly-formed sea ice, helping the animals to feed even during the coldest spells in the depths of winter. The tusks, which are deep-rooted, overgrown, canine teeth, are in turn chisels, adzes, hammers and anchors, which give the animals leverage in hauling themselves about on rough sea ice. Underwater they become ploughs, used in the search for food. Walruses feed on clams, marine worms, fish and other animals which they seek on the sea bed in depths down to sixty metres,

hunting for spells of five to ten minutes between breaths. The tusks stir the ground, the moustache senses the presence of food, and the thick but mobile lips, with surprising delicacy, suck clams from their shells, worms from their tubes, and flesh from the bones of fishes. Walruses move in family groups, usually of one bull, two or three cows, and several calves, including some from earlier matings which have not yet reached independence. Unmated males form large herds or clubs; they hunt together, sleep in malodorous conclave on the beaches, and quarrel unceasingly among themselves in rumbling slow-motion. Even the youngest and smallest walruses seem permanently ill-tempered.

Walruses winter on land or sea ice, usually toward the southern edge of their range, in places where currents and tides keep the sea ice-free. In April and May they begin to move northward, following leads in the sea ice. Calves are born in May, and cows without calves are mated. By June most walruses are in shallow coastal water, diving from ice floes and sleeping off their huge meals in round-the-clock sunshine. There is a southward migration in autumn as the sea begins to freeze over; the walruses avoid patches of heavy ice where leads are scarce, and keep to shallow water within their diving capabilities.

Sea lions and fur seals are included in the family Otariidae—the eared seals. Distinguishing characters are their tiny external ear, and testes contained in a scrotal sac; walruses and phocids have no external ear, and their testes are internal—in either the abdomen or the body wall muscles. Otariid seals also have long, elegant finger nails on

Steller's sea lion of the north Pacific Ocean, named after the naturalist who accompanied Vitus Bering's expedition of 1740. Sea lions have a hair coat without the fur seal's velvety underfur.

Right:
Alaska fur seals. Cows and pups in a Pribilov Island colony. Fur seals and sea lions are well adapted to life in cold water. They seldom live in extreme cold because the long hair of their coats tends to ice up when they emerge from the water.

GILSATER/COLEMAN

Walruses, ungainly on land but elegant in water, make long annual migratory journeys following the retreating ice edge north in summer and returning south in winter. Bristles, tusks and muscular lips are highly functional, helping an epicurean feeder to select his molluscs from the sea bed.

the dorsal surface of their fore-flippers. Sea lions have a blunt, bear-like nose and a hairy coat with little or no underfur. Fur seals have a pointed, dog-like nose, often with long sensitive bristles, and a hairy coat with soft velvety underfur. Sea lions and fur seals probably evolved in the north Pacific, so it is not surprising to find one species of each in that region today. Their hairy coat seems less efficient at very low temperatures than the shorter, stiffer coat of phocids, and both species avoid situations where they are likely to become iced up. Thus they have never spread from the Bering Strait region to other parts of the polar basin.

The Arctic sea lion is named Steller's sea lion, after a naturalist who made many discoveries in the north Pacific in the mid-eighteenth century. This species normally ranges about the Bering Sea and neighbouring coasts, but enters our region at the Pribilov Islands, off the western shore of Alaska. Adult males are formidable animals, over two and a half metres long and weighing up to a ton. Females are only a quarter of the weight; this dimorphism is characteristic of seals and other animals which breed in harem groups, with dominant males fighting to keep control over several females and their offspring. The males take up breeding territories in May, and maintain harems of ten to fifteen cows. Mating follows in June, but the fertilized ovum does not begin to develop until October, and the pups are born in the following early June. Thus, the single social gathering about midsummer ensures that the pups are born within the protection of a harem, and that every cow is made fertile for the following year, though gestation takes only eight months. After breeding, males from the Aleutian Islands swim northward for a late summer vacation in the Bering Strait, moving south again before the ice starts to form.

The Arctic fur seal is the Pribilov fur seal, which breeds on the Pribilov Islands, Komandorskiye Islands, and Robben Island in the Okhotsk Sea. It ranges widely about the northern Pacific Ocean between breeding seasons, visiting California and Japan in winter and moving back toward the breeding grounds in spring. Males are about two metres long and weigh 500 to 600 pounds; females are very much smaller, weighing about 120 pounds. Breeding begins in June, the males taking up territories on rocky beaches and rounding up harems of forty to fifty cows. As with sea lions, mating follows closely after pupping, but implantation and development of the ovum are delayed to

Common seal. Length 1½ to 2 metres. Widespread in Europe and North America as well as in the Arctic.

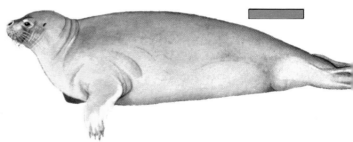

Bearded seal. Length 2 to 3 metres. The moustache probably helps in finding food on the sea bed. Widespread in Arctic waters.

Ringed seal. Length 1½ to 2 metres. Lives and feeds in surface waters among the pack ice of the Arctic basin.

Harp seal. Length 1½ to 2½ metres. Widespread from Baffin – Hudson Bay area to Spitzbergen – Novaya Zemlya.

Ribbon seal. Length 1½ to 2 metres. Found only in the Bering Sea and neighbouring Arctic Ocean; almost entirely on sea ice.

Hooded seal. Length 2 to 3 metres. Large seals of the sea ice off Labrador, Greenland and Jan Mayen.

Grey seal. Length 2 to 2½ metres. Common on both sides of the Atlantic and on rocky shores in southern and western Iceland.

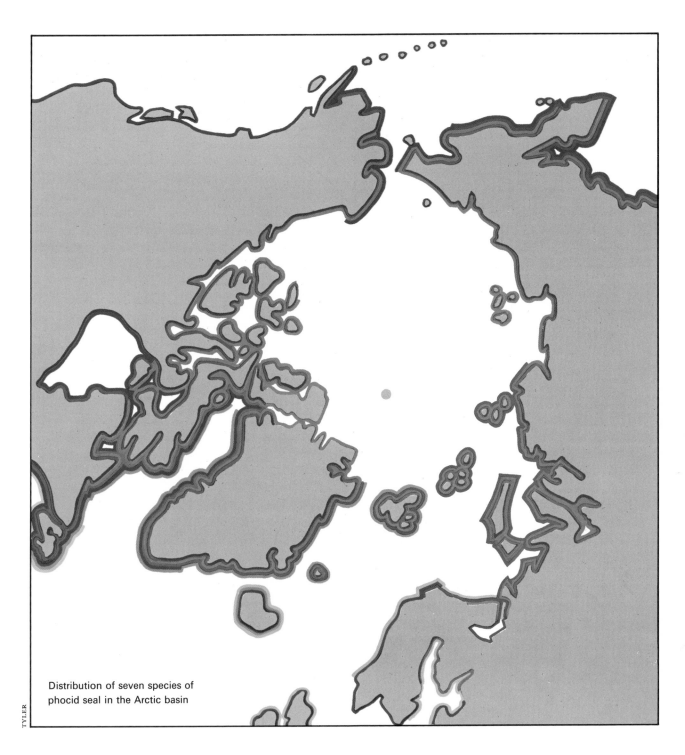

Distribution of seven species of
phocid seal in the Arctic basin

ensure that the pups are born at the right time of year. Pribilov fur seals feed on fish, which they catch both at depth and in surface waters.

This species has been hunted by man for nearly 200 years; its fur is valued above that of any other seal, and many fortunes have been made at its expense. However, since 1911 its exploitation has been controlled sanely by international agreement; man has harvested, rather than hunted the Pribilov fur seal to his own lasting benefit.

The seven Arctic species of phocid seal differ widely in their patterns of distribution. Bearded and ringed seals are circumpolar and based mainly on the polar coast; their home is the fast ice of the Arctic basin, extending southward down the cold shores of Labrador, Greenland and eastern Asia. Common or harbor seals are also circumpolar, but generally live in cold temperate regions and reach the Arctic only at the northern edges of their range. Grey seals are entirely Atlantic animals with Arctic representatives in Labrador, Iceland and northern Europe. Hooded and harp seals are also

Eskimo at seal hole. Breathing holes, kept open by seals throughout the winter, are watched carefully for signs of activity. The seal is harpooned on coming up to breathe. Polar bears also watch seal holes, killing seals with a blow and hauling them up through the hole and onto the ice.

Pups are born as the ice begins to open in April and May. Ringed seals are smaller, measuring one and a half metres and weighing up to 200 pounds. Their coat is distinctively marked with white circles. Though solitary like the bearded seals, they are more numerous and often form small groups on the ice. They feed pelagically on plankton and fishes, probably diving deeply for their food when the plankton descends in winter. Though seldom found on the pack ice, they move out from the coast when the fast ice disperses in spring and make their way back in time for the autumn freeze-up. Their pups are born in March or April, usually in the shelter of a snow cavern, which the mother enters directly from the sea.

Both bearded and ringed seals are hunted by Eskimos, from kayaks in summer and by harpooning at their breathing holes in winter. Apart from the food and oil which they yield, bearded seals are prized for their thick hide, which makes thongs and rope, ringed seals for their soft pelts, which are made into clothing.

Hooded seals and harp seals avoid coastal ice and live well out to sea on heavier floes. Hooded seals are large grey animals, measuring three metres and weighing up to 900 pounds. Harp seals are only two metres long and half the weight, with a pale grey coat, black face and black, harp-shaped patch across the back and sides. Male hooded seals have an inflatable nose, which is frequently blown up during the breeding season. It alters the animal's profile and provides a resonating chamber to enhance its roar; both effects intimidate adversaries and are perhaps admired by females. Hooded seals form breeding groups when the pups are born, congregating in masses of several thousand animals on stable sea ice; there are two major breeding centres, one off southern Labrador and the other close to Jan Mayen Island. Pairs form in March. The pups are born in late March and April, and mating occurs during the two or three weeks following birth. Then the breeding groups disperse. From Labrador many hooded seals move north to spend May and June fishing off southwestern Greenland. In July they move east and north round Cape Farewell, and congregate to moult on heavy pack ice between Greenland and Iceland in late summer. As winter advances they move southward down the Greenland coast, returning to the breeding grounds in early spring.

Harp seals breed in three main concentrations, two of them close to the hooded seal breeding areas and the third on the ice of the White Sea, east of

Atlantic, but extend northward well into the Arctic basin. Banded or ribbon seals are entirely Pacific seals, restricted to eastern Asia south and west of the Bering Strait. Most of the species show slight evidence of local variation, suggesting that there is little mixing between stocks from different geographical areas.

Bearded seals and ringed seals are the phocids of the Arctic basin. Bearded seals measure about two and a half metres and weigh over 500 pounds. They are reddish brown, with the coarse hair coat typical of their family. A most striking feature is their bushy moustache which, like that of the walrus, is probably used in sensing food on the sea bottom. Bearded seals feed on the sea floor, and are said to use their square-cut flippers as shovels in sorting through the gravel and mud. They are solitary for most of the year, living close to the coast; during summer they haul out onto the beaches, and in winter they dive from the sea ice, making use of permanent leads and polynyas as breathing holes.

Hooded seal mother and pup. Hooded seals are plankton feeders, breeding far out on the sea ice. Off Newfoundland and Iceland their pups are hunted for their sleek grey coats.

northern Scandinavia. White Sea pups are born in late February, Jan Mayen and Newfoundland pups in March and April. Harp seals disperse northward from their breeding areas, moving with the drifting ice and feeding pelagically as they go. From September they begin their southward movement, keeping to patches of open pack ice through the winter. Both hooded and harp seals are hunted commercially. The white, new-born pelt of the harp seal, and the silver-grey skins of first-year pups of both species, are particularly valuable; adults are also taken for their skin, oil and meat. Hunting is controlled by the nations concerned, but little research has been done on these species and there are growing fears of over-exploitation.

Grey and common seals are interlopers from the south which have gained a footing in Arctic regions. Grey seals are long, lithe creatures in comparison with true Arctic species; a large male measuring three metres may weigh only 500 to 600 pounds. Females are slightly smaller than males. The silver-grey coat is distinctive, and grey seals sometimes belie their kinship enough to grow a small external pinna to their ear. They breed on rocky shores, usually in harems. In the western Atlantic population of Newfoundland and Labrador, pups are born in February and March. European and Icelandic grey seals breed mainly in October and November. All keep well clear of ice during their wide dispersal between breeding seasons

Ringed seal. The fawn rings characteristic of the species can be seen on the flanks. This is the commonest and most widespread Arctic seal, providing Eskimos with meat, oil and clothing.

Bull hooded seals inflate their bladder-nose when excited or irritated. The hollow bladder is a resonating chamber and probably a ritual threat in sexual display.

Grey seals sunning. Frequenting rocky shores and coastal waters, this species eats fish and is often unpopular with local fishermen.

Common seals are smaller, weighing up to 250 pounds. Their coat is spotted black on silver-grey, with local variation in the five or six geographical races of the species. They reach Arctic waters in Hudson Bay (where one small population is entirely land-locked in freshwater lakes), in Labrador, Greenland, Iceland, northern Scandinavia, and the Bering Strait region. Over most of their wide range common seals breed in late spring, characteristically producing their pups on sandbanks in tidal water. The pups have a woolly natal fur, which in these southern stocks is shed before they are born; thus they are ready to swim almost from the moment of birth, and certainly ready to take off with the next high tide. Western Pacific stocks of the same species give birth on sea ice in February, March and April. These pups retain their natal coat for the first few weeks of life, as an additional defence against cold, and put off their first swim until they have built up an insulating layer of fat under the skin.

Both grey and common seals eat fish, which they usually catch close to the shore. This brings them into competition with fishermen, who destroy them as competitors rather than for commercial gain. Icelandic common seals are hunted for their silvery yearling coats, and northern Japan has a commercial seal fishery based on the same species.

Ribbon seals are a little-known species of eastern Siberian waters. They are small, weighing up to 200 pounds and measuring less than two metres. Their coat is an unusual chocolate brown, with three-inch-wide creamy bands encircling the neck, flippers and pelvis. They winter on sea ice, producing their pups on floes in March and April. Ribbon seals are believed to feed pelagically and do not appear to migrate. A few are taken each year by commercial fishermen, but there is no great demand for their skins, blubber or meat, and the species seems to be safe in its far eastern haunts.

Polar bears and sea otters

Most bears are terrestrial, and either vegetarian or omnivorous. Polar bears are marine animals which can, if necessary, spend their whole lives without stepping on land, and they are mainly carnivorous. They are among the largest bears. A big male measures two and a half metres from jet-black nose to stubby tail, stands one metre high at the shoulders and higher at the rump, and weighs up to half a ton. They pack tremendous power in the muscles of their jaws, neck and limbs. The huge clawed feet are well adapted for running on ice, and effective in clubbing and tearing prey. Polar bears are the dominant predators of the Arctic, and a match for any other species on land or ice. They swim well, though clumsily in comparison with seals and whales. Practically all of their life is spent at sea, riding the ice floes, lumbering from lead to lead in search of food, and swimming between floes when necessary.

They are solitary animals with little social life, shunning each other on sight and seldom travelling in company or fighting. They mate in March and April, males and females wandering briefly together and sometimes becoming involved in sexual

Harp seal mother with two- or three-weeks-old pup. This species breeds on pack ice off Labrador, Greenland and northern Europe, where pups are slaughtered for their white fur and there is concern for the future of the species.

Right:
Common or harbour seals are widespread in the northern Atlantic and Pacific oceans. Their pelt is prized by Eskimos as decoration and trimming for clothing.

LEE RUE/COLEMAN

Grey seal pups. Grey seals are widespread in the north Atlantic, reaching Arctic waters off Labrador and Iceland.

BREUMMER

69

Eskimos say that, while stalking seals over sea ice, the polar bear hides its black nose with one paw—to help it remain invisible.

Following pages.
Northern gannets. Familiar plunge-divers of north and west Britain, gannets breed in tightly packed colonies on northern Atlantic stacks and cliffs.

Polar bears are strong but clumsy swimmers and wander widely over the sea ice. Walking and swimming, they visit all the northern islands. King Karl's Land, Svalbard.

fights with rivals. Then they part, to continue their restless patrol over the breaking sea ice. Through the summer they travel alone on the open pack ice, moving toward the coast in autumn. As winter approaches, all pregnant females, most other females, and a high proportion of males seek out a lair—a rock shelter on land or an ice cavern on the sea ice, where winter snows will build a mound over them. There they settle and sleep. It is not a true hibernation, for body temperature falls only slightly and the bears can awaken at any moment. Sleeping out the winter in their thick overcoats, and under a blanket of insulating snow, reduces their energy losses to a minimum and allows them to survive the lean months when prey is scarce.

Females give birth to their cubs in January and February. Usually two are born, no larger than half-grown kittens, and they nestle in their mother's fur for the first month or six weeks of life. In March or April they leave the lair and wander with their mother over the ice. They remain with her for two years, slowly learning the skills of polar hunting and ice-craft. Fully grown at last, they leave her as she re-mates and go their separate ways.

Polar bears feed mainly on ringed seals. They catch the seals by stealthy stalking on the ice, or by stunning them at the breathing holes during the winter. Other species are taken as well, though polar bears seldom tackle walruses and go out of their way to avoid them in the water. Fishes and other marine prey are taken, and on land a hungry polar bear will eat anything—including lemmings, berries, eggs and carrion of all kinds. They are hunted extensively by polar man, who relishes their meat and grease, and uses their skins for clothing and bedding. Commercial and trophy killing are in general discouraged by the governments of the northern countries where polar bears wander, and the species does not at present seem to be in danger from man.

Sea otters, which are larger and heavier than the more familiar river otters, live along the Pacific shore of North America from California to the Aleutians and southern Bering Sea. They rummage for clams and sea urchins along the shore and sea bed; bringing shelled food to the surface, they hold it against their chest and crack it with rocks. Sea otters keep company with each other in the water and play in groups around fishing boats. Easily netted or shot, they have been reduced almost to extinction by demand for their fine, silky fur, and are now protected. Females produce only a single cub each year, usually in the water, so the recovery rate of the species is likely to be slow.

Sea Birds of the Arctic

While seals, whales and fishes play havoc with the plankton from below, sea birds by the thousand attack it from above. Arctic and Subarctic waters are the haunt of about fifty species of maritime birds, many of which are present in enormous numbers. Some, like the petrels and alcids (auk-like birds), feed entirely at sea throughout the year. Others—gulls, terns and skuas, for example—feed at sea in winter, but come ashore in summer to take what the land can offer. Some feed at sea while they are waiting for winter snow to disappear and lakes to open; phalaropes, divers and many ducks fall into this category. Whatever their dealings with the sea, they generally feed at the surface, or close to the bottom in shallow water. They help to keep nutrient salts in circulation, and accelerate the flow of energy by their steady browsing on the plankton. Finally, they are responsible for carrying enormous quantities of nutrient salts ashore in their droppings. Bird colonies in current use, and the sites of abandoned colonies, are always oases of nitrogen-enriched soil and lush vegetation.

Though practically all of the Arctic sea birds feed on plankton, they catch it in a variety of ways. Each species has its own preferred food among planktonic animals, usually relating to the size and shape of bill, the depth to which the bird is capable of diving, and other morphological or physiological factors. So the dozen-or-so species of sea birds feeding together in a shoal of plankton are more likely to be dividing the food between them than competing. Though several may be concentrating on two or three of the most prominent planktonic species, they may be taking different age groups, hunting at different levels, or in other ways contriving to share the loot without direct competition.

Petrels

In Antarctic waters two dozen species of petrels breed and feed; as many as a dozen species may sometimes be seen feeding together in mixed flocks. Arctic waters have only three breeding species of petrel, and only two other species which make regular visits. The ecological role of the missing species is taken by alcids, who also double for penguins in the Arctic Ocean. The most prominent breeding species is the fulmar, a large, heavy-billed petrel with wings spanning over a metre. Fulmars breed in Iceland, Svalbard, Greenland, the Canadian archipelago and the Siberian polar islands. During the past century their numbers in temperate

Arctic petrels. Petrels divide the rich surface waters of the ocean between them. Fulmars settle and bob for their food, storm petrels scamper over the surface and dabble elegantly, shearwaters dive and swim completely submerged after active prey.

Leach's storm petrel breeds mainly in the cool temperate zone, but reaches the Arctic fringe in Iceland and Labrador.

Like a heavily built gull or a very small albatross, the fulmar is a splendid glider and flies powerfully against the wind. Like all other petrels, it lays a single, large egg.

Shags nest close to the sea-shore using seaweed, sticks and grass as nesting material. Their main food is fish which they chase under water, propelled by strong legs and feet.

waters, possibly also in the Arctic, have increased enormously. They are gull-shaped, with narrow gliding wings and the tubular nostril which is the hall-mark of petrels. In the far north most fulmars are brown. At southern colonies in Iceland and Britain they are mostly grey and white, and the proportions of the two forms vary with latitude between these extremes.

The fulmars of Jan Mayen and Bear Island return to their cliff colonies in midwinter, and are able to feed among the broken ice close to the islands. Elsewhere, they take up the nesting ledges in April or May, as the winter ice begins to break up. The large, single eggs are laid in June, and the chicks leave the nests in September. Fulmars glide gracefully over the water like miniature albatrosses, feeding mainly on small fishes and crustaceans which they catch by bobbing from the surface. They are also scavengers, following the fishing boats to feed on offal from the catches. The recent increase in fulmar numbers has been related by some to the spread of commercial fishing activities in the northern Atlantic during the same period. The present trend toward processing offal at sea, rather than throwing it overboard, could lead to an equally rapid decline in numbers during the next few years.

The remaining breeding species are the Atlantic storm petrel and Leach's storm petrel, two very similar little birds of temperate seas which enter the Arctic region in Iceland. They breed in burrows on rocky islets, and feed typically by dancing over the waves, pecking busily at the water surface as it curls about them. Two other species visit the Arctic from the south. Manx shearwaters spread each autumn from breeding grounds in Britain and southern Europe, taking in the Arctic fringe as part of a circum-Atlantic tour. Sooty shearwaters breed on islands in the temperate zone of the southern hemisphere. Southern Pacific stocks fly north to Bering Strait, Atlantic stocks to Iceland, during the northern summer. Shearwaters dive and swim for their food within a few inches of the surface, using their folded wings as hydroplanes and their strong, webbed feet as propellers.

Gannets and cormorants

Northern gannets breed in noisy, jam-packed colonies off Newfoundland, Labrador, Iceland and Britain. They usually nest on an island or stack to avoid the attention of ground predators, and build scruffy nests of sticks, seaweed and bones on the flat top and broad ledges, leaving narrower ledges to gulls and alcids. They lay a single egg in May, flying south with their chicks before the cold weather begins in autumn. Gannets feed by plunging from a height into the water, usually catching fish one by one. They keep clear of pack ice, but dive in very cold water off southern Greenland and Baffin Island in summer.

Common or black cormorants, and green cormorants or shags are two other species of diving birds which extend from warmer latitudes to the Arctic fringe. Common cormorants are all-black birds of a widespread cosmopolitan species, which have secured a breeding niche in southern Greenland, Iceland and Labrador. They dive in coastal waters, seldom feeding far out to sea. Shags are very similar but slightly smaller birds, which dive in shallower water and take smaller food. They range from the Mediterranean to Iceland and northern Scandinavia. Both species make prolonged diving sorties, chasing their prey actively through surface waters or along the sea bed. To reduce buoyancy they wet their plumage before diving, and are often seen drying it out in the sun after feeding. They lay three or four eggs in untidy, stick nests, usually in foul-smelling cliff colonies which they share with other species.

Phalaropes

Grey and red-necked phalaropes are wading birds which, like many other waders, nest on the Arctic tundra in summer and feed locally in ponds and lakes while breeding. Between breeding, when most other waders become shore birds, phalaropes turn into deep-water sailors and feed in mid-ocean. Grey phalaropes nest in Alaska, Canada, Baffin Island, Greenland, Iceland, northern Scandinavia and Siberia, and on several islands of the polar ocean. Red-necked phalaropes have a rather more southerly breeding range, though the two species overlap considerably.

Both lay in May or June, producing four eggs. The females are vividly coloured, and take no further interest in the eggs after laying; the males are sombre in plumage, and take all responsibility for incubating, brooding and rearing the chicks. In summer the birds feed on insects and fresh-water plankton in the streams and standing water of the tundra. They catch their food by spinning rapidly in a tight circle, forming a little whirlpool which disorients living prey and concentrates small particles. In winter they fly far south into the Atlantic

BREUMMER

Common or black cormorants. Larger and blacker than shags, they feed offshore in spring and summer, moving south in autumn as the sea freezes over.

TULLOCH/COLEMAN

Red-necked phalarope. In winter these are oceanic sea birds, feeding far south in the Atlantic and Pacific oceans. In summer they are breeding birds of the Arctic fresh water, feeding on aquatic insects.

Kittiwakes. Graceful gulls, with behaviour adapted to breeding in cliff colonies. The chicks remain fast on their narrow ledges until fully ready to fly.

BREUMMER

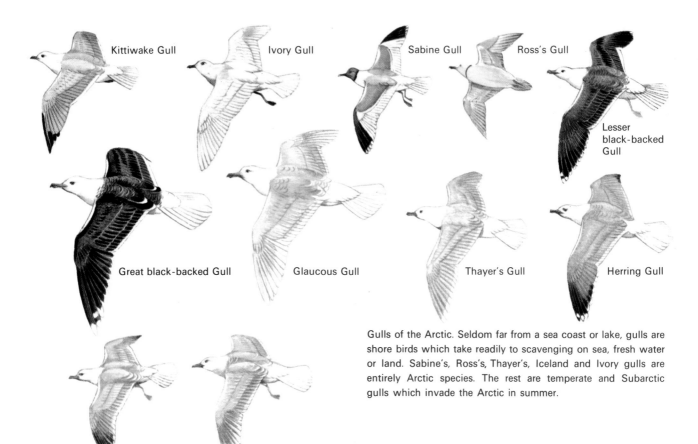

Kittiwake Gull

Ivory Gull

Sabine Gull

Ross's Gull

Lesser black-backed Gull

Great black-backed Gull

Glaucous Gull

Thayer's Gull

Herring Gull

Common or Mew Gull

Iceland Gull

COOK

Gulls of the Arctic. Seldom far from a sea coast or lake, gulls are shore birds which take readily to scavenging on sea, fresh water or land. Sabine's, Ross's, Thayer's, Iceland and Ivory gulls are entirely Arctic species. The rest are temperate and Subarctic gulls which invade the Arctic in summer.

and Pacific Oceans. Immediately before and after breeding they spin and dip for food among the inshore pack ice of the polar ocean.

Gulls, terns and skuas

These form a closely-knit group of coastal and oceanic sea birds with many Arctic representatives. Gulls are intensely social birds, mentally attuned to watching each other closely and feeding communally. These are excellent behaviour patterns for species whose food supply is usually generous but patchily distributed. Though primarily coastal scavengers and inshore plankton feeders, they are equally at home on grassland and in fresh-water lakes and ponds. They have adapted well to agriculture, following the plough and taking every opportunity to feed when man opens the ground for them. This opportunist attitude to life has brought them success in the Arctic, where a succession of different foods may appear throughout the season, and nobody can afford to miss a good meal. When one source fails, they spread out, search for another, and their flocks converge on the fortunate individuals who strike lucky.

Association with man has brought further success;

gulls scrounge from his fishing vessels, scavenge in his back yard, turn over his rubbish tips, and generally profit from his extraordinarily wasteful methods of feeding. Recent climatic changes have favoured them further, generally promoting a northward shift of populations. Thus in Iceland glaucous gulls have been replaced within living memory by greater black-backed gulls from the south, and common and black-headed gulls have spread steadily northward from western Europe into Iceland and the northernmost tip of Scandinavia.

Kittiwakes are oceanic gulls, feeding almost entirely on plankton in summer and winter. They nest on high cliffs overlooking the sea. In the Arctic they are found wherever open water occurs locally in spring and early summer, even on the northernmost islands of the polar ocean. Kittiwakes gather at the breeding grounds in April or early May, following leads in the opening sea ice, and begin laying in late May and June. The two chicks of each nest are fledged by late August, when the whole of the breeding populations in high latitudes begin to move southward toward the warmer oceans. They winter pelagically across the length and breadth of the northern oceans.

Ivory gulls are the most completely Arctic of all gulls, breeding in small colonies on cliffs and rocky coastal plains in northern Greenland and the Canadian and Siberian polar islands. They are pure white, with dark legs and a short, yellowish bill. After wintering at the southern edge of the pack ice, they move northward in May and June, to begin breeding in late June or July. They feed partly on plankton, which they pick from the surface without settling, and are accomplished scavengers where other animals have fed. Ivory gulls accompany man and other predators in their travels, taking scraps, carrion, faeces and any other materials likely to contain digestible protein or fat. After breeding they return to the sea, to forage between the ice floes and follow bears, foxes and other animals whose activities might result in a meal.

Sabine's gull, with grey head and mantle and brown bill, breeds on the tundra and Arctic coast and on most of the polar islands. It feeds mainly on tundra insects and spiders in summer, and winters pelagically in the northern oceans. Ross's gull, one of the rarest of gulls, is a small, pink bird with a thin collar-band. It is very thinly distributed over the high Arctic tundra, feeding in fresh-water lakes in summer and over the oceans in winter.

Glaucous gulls and great black-backed gulls are a closely allied pair of scoundrels; they are the world's largest gulls, with a wing span of over one and a half metres. Glaucous gulls are grey-mantled. They breed on cliffs throughout the Arctic region from Alaska to Siberia, as far south as Iceland and Labrador. Great black-backed gulls have a jet-black mantle. They breed only in the Atlantic region, overlapping with glaucous gulls in Labrador, southwest Greenland, Iceland, Svalbard, Bear Island and northern Scandinavia. They nest on flat or rolling ground rather than cliffs, generally in solitary pairs or small groups. Both species are predatory, killing other birds on the wing and stealing eggs and young. They also scavenge after hunting animals, and generally live by their sharp wits during the summer season of plenty. In winter they move south to feed on plankton at the edge of the pack ice.

Herring, lesser black-backed, Thayer's and Iceland gulls are four species of medium size gulls, closely akin to each other and very similar in appearance and ecology. Herring gulls, the commonest gulls of the Atlantic seaboard, have a blue-grey mantle, black flight feathers with white wing tips, and pinkish legs. Widespread in Canada, this species is believed to have spread to western

BURTON/COLEMAN

Europe fairly recently, perhaps in the last few thousand years. Certainly in the past half-century they have spread northward into Iceland, Bear Island, Svalbard and northern Scandinavia. They nest on cliffs and rolling ground throughout the Canadian Arctic and along the European coast. Lesser black-backed gulls are the European native equivalent of herring gulls, slightly smaller, with dark mantle and (usually) vivid yellow legs. In northwest Europe they seem gradually to be retreating before the continuing spread of herring gulls, and they too are spreading northward into the Arctic. They form large colonies, usually on flat ground, and their Arctic range includes Iceland, northern Scandinavia, Russia, Siberia, and the Siberian Arctic islands. Polar populations of herring and lesser black-backed gulls move south-

Glaucous gull, one of the two largest species of gulls, displaying its metre-and-a-half wing span. This species feeds mainly on land, preying on other birds and scavenging.

Left: A colonial gull common on farmland in western Europe, the black-headed gull has spread to Iceland and northern Scandinavia in recent decades.

Herring gulls, originally North American, have recently spread widely along the eastern Atlantic seaboard and in the Arctic. They benefit from commercial fishing, following the boats and feeding on scraps.

Great black-backed gull, an Atlantic species that breeds as far north as Svalbard and Bear Island in the Arctic. Like Glaucous gulls, these birds are predators and scavengers.

Following pages: Kittiwakes range far over the ocean in winter. In spring and summer they feed closer to home, bringing small fish and other planktonic foods to their growing young.

Lesser black-backed gulls, darker and smaller than herring gulls, are spreading into the Arctic from northern Europe. Climatic improvement and increasing activity by man is probably responsible for their spread.

ward after breeding to winter along warmer coasts of America and Europe. They eat practically anything, including plankton, intertidal animals, carrion, and each other's eggs and chicks.

Thayer's gull is a local variant of the herring gull, breeding along the western flank of Greenland, throughout the southern islands of the Canadian archipelago, and on the Canadian mainland tundra coast. They generally have a pale brown eye and grey flight feathers. In winter Thayer's gull moves westward to the Pacific coast of America. Iceland gulls are yet another variant, with silvery grey mantle and flight feathers. They breed on Ungava Peninsula, the southern ends of Baffin Island and Greenland, Iceland, Svalbard, and northern Scandinavia, nesting on cliffs overlooking the sea. They feed pelagically and do not move far from their breeding grounds in winter.

Common or mew gulls are similar again to herring gulls, though smaller and with a distinctly greenish tinge to their bill and legs. They are perhaps the most versatile of all gulls, breeding on coasts, along rivers of central Europe and Asia, and on lakes in the middle of the Canadian forest belt. Originally Eurasian, they spread to Canada via the Bering Strait during the present interglacial or

First-year (front) and second-year immature Icelandic gulls. Most gulls have speckled plumage up to their third or fourth year; they do not breed until they attain full adult plumage.

Great skua or bonxie, fisherman, predator and pirate of the north Atlantic. Fierce in defence of nest and territory during the breeding season, great skuas tolerate each other's company more readily when breeding is over for the year.

Left, below: Arctic jaegers breed throughout the Arctic, feeding on fish, small mammals and birds. They disperse over the oceans of both hemispheres in winter, rarely coming ashore.

Arctic terns breed in temperate as well as Arctic regions, usually in large colonies on grass or shingle. Versatile flyers, they pick their food from the surface of sea or fresh water.

possibly earlier, and they breed far north along the Mackenzie River; in Europe they reach the Arctic around the White Sea coast and northern Scandinavia. Their numbers seem to be increasing around Iceland, where breeding began only a few years ago. Like most of their kin, common gulls feed happily on anything they can find. Being smaller than most other gulls, they can to some extent avoid competition by taking smaller food species, or concentrating on kinds of food which the larger gulls are too clumsy to catch. On the tundra they eat mice and voles, insects, and small aquatic invertebrates from the lakes and streams. In winter they become coastal scavengers, poking with their slim but powerful bills into crevices which other gulls cannot negotiate, and digging into the soil in search of earthworms and insect larvae.

Terns are delicate, slender-billed gulls with narrow, pointed wings and tail. Adapted for speed and highly manoeuvrable flight, they dart over the ocean surface in search of small fish and plankton, dipping briefly but hardly wetting their feathers. They are equally at home over fresh-water, taking insects on the wing and larvae from the lake surface. Of the many northern hemisphere terns only one breeds north of the tree line. The Arctic tern nests colonially on rolling tundra, marsh or shingle. Its Arctic range extends across northern Alaska and Canada, Greenland, Iceland and Arctic Eurasia, and to most of the polar islands. The soft grey mantle, translucent white wings and tail, coral bill and black cap make it one of the most attractive Arctic birds. It is also one of the busiest. Arctic terns arrive on the breeding grounds in early June and waste no time in laying and incubating. The eggs, beautifully camouflaged, are laid directly on warm shingle or dry vegetation, and intending predators are attacked fiercely by every bird in the colony. By August or early September both fledglings and parents are starting to leave. They fly southward in small flocks down the length of the Atlantic and Pacific Oceans into the southern hemisphere, some as far as the Antarctic pack ice and continental coast. This is one of the longest migratory journeys known among birds, and a surprising feat for so flimsy and delicate a creature.

Skuas and jaegers are brown, gull-like birds which feed ashore while breeding, but are completely pelagic for the rest of the year. They are fiercely predatory and breed in solitary pairs, though usually grouped in loose communities. Great skuas, the largest species, are widely distributed in the temperate and polar southern hemi-

LEE RUE/COLEMAN

Long-tailed jaeger. Like all other skuas and jaegers, this species is a predator of small mammals as well as an accomplished fisherman.

TULLOCH/COLEMAN

BREUMMER

Brunnich's guillemot, an old-world species which breeds on cliffs of the western North Atlantic and Siberian Arctic and feeds among the pack ice throughout the year.

Left: Black guillemots breed on either side of the north Atlantic, in Hudson Bay, and throughout the Arctic from Alaska to Siberia.

Cliff colony of common guillemots. Several in the foreground are "bridled"; the white eye-ring and streak have no known function, but appear much more frequently in northern than in south common guillemots.

sphere. In the north they breed only on Iceland, the Orkney and Shetland islands, and the Faroes. Locally known as "bonxies", great skuas are stolid ruffians with a taste for fish, plankton, small birds and carrion. Some food they catch honestly for themselves at sea. The rest they steal, bullying home-coming sea birds, which are frightened into disgorging the food intended for their young, robbing nests, and knocking small birds ruthlessly to the ground. The jaegers are slender, piratical skuas with a similar mode of life. They breed entirely in the northern hemisphere. The three species—pomarine, Arctic and long-tailed jaegers —have a very similar distribution across the tundra of Canada, Alaska, Eurasia, Greenland and the Arctic islands. Pomarine and long-tailed jaegers are found wherever there are voles and lemmings

for them to feed on. Arctic jaegers are less dependent on small mammals and breed in places where mammals are scarce or absent, for instance in Svalbard and Iceland. Both skuas and jaegers nest on the ground, laying two or three eggs and rearing families of voracious, piratical young. The breeding success of pomarine and long-tailed jaegers varies greatly from year to year, depending on cyclical fluctuations in the populations of their prey species (page 118). In winter, skuas and jaegers spread southward from the breeding areas and feed entirely over the ocean. Some fly far into the southern hemisphere, to be recorded occasionally in coastal waters of South America and Australasia.

Alcids

About fourteen species of alcids breed in Arctic waters, including auks, guillemots, razorbills, puffins, and diminutive auklets and murrelets. They are compact diving birds, often compared with penguins. Their small, narrow wings are used both in flying and for swimming under water. Practically all of their food comes from the sea and is caught by diving. The wing is too thin and small

to allow gliding or hovering, and alcids have no ability as scavengers. They fly easily and for long distances with very rapid wing beats, always at high speed, often planing for miles over the surface of the water like hydrofoils. The larger species nest mainly on cliffs and rocky islets, laying a single pointed egg on the bare, rock ledge. Smaller species usually burrow, laying their eggs several feet underground. Large alcids often dive deeply, taking food at the bottom in depths of over 100 metres. Smaller species take more of their food from the surface.

Of the four species of guillemot found in Arctic waters, Brunnich's guillemot and the smaller black and pigeon guillemots range furthest north. They breed on many islands of the polar basin, Brunnich's and black guillemots in Iceland, Greenland, and Siberia, pigeon guillemots replacing black guillemots in the Bering Strait area. Black guillemots also breed in Scandinavia, Labrador and Hudson Bay. Common guillemots generally range further south, though they reach the Arctic in Labrador, Iceland, northern Scandinavia, Greenland and Novaya Zemlya. Between breeding seasons most northern guillemots move away from their breeding cliffs to fish in open water among the pack ice.

Razorbills and puffins are north Atlantic birds

Razorbills. These were the original "pingouins" of which Anatole France wrote in his political satire *Island of Penguins*. Southern-hemisphere penguins, although completely unrelated, were named after them.

Right: Puffin. The most colourful of the Arctic seabirds.

FATRAS/JACAN

Tufted puffins. One of the many species of small alcids which feed on the plankton of the northern Pacific Ocean.

The huge bill casing of the puffin is brightest during the breeding season, losing some of its colour afterwards. Like other alcids, puffins of the far north are important additions to Eskimo diet.

with breeding grounds in Labrador, Newfoundland, Greenland, Iceland, Jan Mayen, Bear Island and Scandinavia; in addition puffins reach Novaya Zemlya and Svalbard. Razorbills are similar to guillemots, with a heavier, white-striped bill. Puffins are smaller, with a huge, parrot-like bill which takes on a variety of colours in the breeding season. In the north Pacific Ocean horned and tufted puffins reach the Arctic on islands in the Bering Sea and cliffs close to the Bering Strait. Razorbills and the various kinds of puffins dive deeply for some of their food, but also catch fishes in surface shoals.

The smaller species of alcid include the dovekie or little auk of the northern Atlantic, and the tiny auklets and murrelets of the north Pacific. Dovekies are engaging small birds, like very fat thrushes with bumble bee wings, which breed in crowded tenement cliff colonies. They are concentrated in Greenland and on the polar islands north of Siberia. Many colonies numbering hundreds of thousands are known. Dovekies feed on plankton, plummeting into the waves and emerging with wings whirring steadily. In winter they disperse south of the breeding areas, feeding in huge, scattered flocks across the northern Atlantic Ocean. In the north Pacific a

similar role is played by least, crested and parakeet auklets and Kittlitz's murrelet, which breed and feed in close company in the Bering Strait region, Bering Sea and Aleutian Islands. On the bare, talus slopes of the breeding islands, the three species of auklet select nest sites according to size. Where the rocks and the cavities between them are small, least auklets predominate. Crested auklets, three times as big, nest among large boulders, and parakeet auklets nest less gregariously among the pinnacles and ridges of the islands. Murrelets lay on the surface of stable talus slopes, so avoiding competition altogether. The auklets similarly divide their planktonic food between them, each species concentrating on the range of crustaceans best suited to the size of its bill.

The great auk or garefowl, largest of all the alcids, at one time nested on offshore islands from southern Labrador to Greenland, Iceland and northern Britain. With the shape of a very big razorbill and completely flightless, it wintered as far south as Florida and the Mediterranean coast. Systematic killing and nest-robbing by passing sailors brought the species to the verge of extinction early in the nineteenth century; the last great auks were killed in 1844.

PUCHALSKI/COLEMAN

Little auk or dovekie, smallest of the Atlantic alcids.

Arctic alcids. Sometimes confused with penguins, the alcids are diving birds which are also strong fliers. Many nest on high cliffs out of reach of predators, and fly long distances over the sea in search of food. Great auks, too heavy to fly, were destroyed by man in the 18th and 19th centuries.

Great Auk

Parakeet Auklet

Horned Puffin

Pigeon Guillemot

Kittlitz Murrelet

Crested Auklet

Least Auklet

COOK

Black-throated diver at its lake-side nest. Divers nest in solitude close to the water's edge, sliding on and off the nest unobtrusively. The nest is lined with rotting vegetation.

Red-throated diver. The band of dark chestnut on the throat looks black from a distance. Plumage patterns on the back and flanks distinguish this species from the black-throated diver.

Common or great northern diver. The divers or loons are the most efficient of all diving birds; adapted for swimming, with legs far back along the body, they cannot walk upright on land.

Divers

The divers or loons are a family of large, solitary water birds, which breed on tundra lakes and winter at sea. They are the most efficient and highly adapted of all diving birds, capable of staying underwater for two to three minutes, and of diving in depths of sixty metres or more. Of the four living species, one—the great northern diver—breeds mainly on forest lakes of North America and in Greenland, Iceland, Jan Mayen and Bear Island. The remaining three species—the white-billed, black-throated and red-throated divers—breed in northern Eurasia as well as North America, both on tundra and on forest lakes. The red-throated diver also breeds far to the north in Greenland, Ellesmere Island and the islands north of Siberia.

All four divers have a similar ecology and way of life. For about three months each year they occupy a lake territory, nesting on the shore or in floating vegetation, raising a brood of two chicks and feeding mainly on fish. Red-throated divers usually select small lakes, which lose their ice early and warm in the sun; black-throated divers prefer cold, deep lakes, and white-billed divers are often

Arctic waders. Waders or shorebirds spend half their life on estuaries and coasts, feeding in the intertidal zone. Each summer the Arctic species fly north to the tundra, where they feed mainly on insects of grassland and shallow water. The different shapes and proportions of bill allow many species to feed alongside each other, taking different foods and so sharing the available resources. Waders are described more fully under "Birds of the tundra" on pages 144–149.

Grey Plover

European Golden Plover

American Golden Plover

Semi-palmated Plover (Ringed Plover)

Whimbrel (Hudsonian Curlew)

Turnstone

Dotterel

Curlew

Dowitcher

Bar-tailed Godwit

Knot

Purple Sandpiper

Sanderling

Dunlin

Black-tailed Godwit

Ruff

CASSELLI

Common Eider

King Eider

Spectacled Eider

Steller's Eider

Harlequin

Goosander

Goldeneye

Red-breasted
Merganser

Common Scoter

Velvet Scoter

Oldsquaw

COOK

Arctic sea ducks. Sea ducks are mainly diving ducks, adapted for feeding on or near the bottom in shallow coastal water, lakes and rivers. They usually winter in coastal waters, and breed in summer along the coast or lake edge. The largest tend to dive deepest, leaving the shallows for smaller species.

Eider ducks line their nests with "eider down", one of the most efficient natural insulators known to man. They nest close to open water, leading their chicks for their first swim as soon as the eggs hatch.

found on marshes and river deltas. From September they move with their fledglings to the shore, black-throated and red-throated divers often flying south to winter on temperate or subtropical coasts. Great northern and white-billed divers rarely stray far from their breeding areas and are quite happy to winter among floating ice.

Sea ducks

Of the twenty-odd species of ducks which breed in the Arctic region about half spend part of their life at sea. Some are almost entirely marine, breeding and wintering along the shore. Others breed inland on rivers or lakes but winter close to the sea. In the first group are the four species of eider duck. Common eiders breed along both shores of the Bering Strait, the northern Canadian coast, Labrador, Newfoundland, Baffin and Ellesmere Islands, Greenland, Iceland, Scandinavia, eastern Siberia, and many of the Siberian polar islands. King eiders have a similar but generally more northern distribution; they do not breed in northwestern Europe, Iceland or southern Greenland. Spectacled eiders and Steller's eiders breed

in the Bering Strait area. Eiders usually nest in large, scattered colonies along the coast, often where a lake or lagoon provides fresh water close at hand. Man takes an interest in the welfare of eider ducks, providing stone shelters for them to nest in, warding off predators, and encouraging breeding in every way possible. Eider down, the soft breast plumage which the bird plucks to line her nest, is one of the lightest and most efficient thermal insulators known, and much in demand for bedding.

Many other species of duck breed close to tundra lakes, diving in fresh water and rearing their chicks on the fishes, insect larvae and molluscs of the lake habitat. At the first signs of approaching autumn they leave the tundra and fly to the coast, usually making their way southward to ice-free or even temperate waters. Harlequin ducks winter in cold waters off Iceland and Greenland. Velvet and common scoters head south into the Atlantic and Pacific Oceans, reaching latitudes between 30°N. and 40°N. Oldsquaws (long-tailed ducks), red-breasted mergansers, goosanders and goldeneyes winter in coastal waters or large rivers, estuaries and lakes.

Life on the Tundra

Glacial valley recently exposed by the retreat of its ice sheet. The foreground, longest exposed, has had time to develop a mixed flora. The most recently exposed ground (distant) is still relatively bare.

Previous pages:
Caribou herd (centre) crossing the Canadian tundra.

Constant freezing and thawing churns the soil, pushing the larger stones into piles and separating the finer materials. Sorted soil moves too quickly for plants to become established.

Tundra is a Lapp word meaning the vast, rolling, treeless plain of northern Europe. In ecology it describes the kind of environment (including soil structure, topography and vegetation) found above the tree line, either at sea level in polar regions or on high ground further south. Arctic tundra merges into alpine tundra, and alpine tundra may extend at increasing altitude from polar regions to the tropics. Tundra is new ground, which, for one reason or another, has not had time to acquire complex, tightly-knit communities of plants and animals. Some is new because the ground has only recently been released from its covering of ice—much of the Arctic has been ice-free for only tens or hundreds of years. Some is new because the ground surface is constantly being renewed by violent weathering—mostly mechanical abrasion and frost action. Practically all of the Arctic tundra is new in this respect. One indication of its newness is the immaturity of its soils.

Arctic soils

Soils begin to form when rocks break down mechanically and chemically, by the various forms of weathering and erosion. Re-sorted by wind and water, the breakdown products yield a matrix of fine clay and silt particles which are a primitive, *ahumic soil*. Ahumic soils are often too acid or too alkaline, too dry or too likely to be blown away, to make a permanent home for plants. But their interstices hold water, which allow simple plants to find a footing. They are the typical soils of the coldest, driest or newest parts of the tundra, with a flora of bacteria or algae but very little else.

Plant cells which manage to settle add organic material to ahumic soils, slowly converting them toward *humic* or *organic* soils. These are more than a mixture of rock dust and dead vegetation. Chemicals released by the plant material react with the finely-divided rock to release minerals, some of which stimulate plant growth. Blue-green algae and nitrogen-fixing bacteria begin to settle. These are important plants which extract nitrogen from the air and release it in solution for the use of other plants. Organic buffers develop, which neutralize the excess acidity or alkalinity and provide a more attractive chemical environment for plants and small animals. The growing mat of vegetation holds water, shields the soil from sun and frost, and helps to create a stable, physical environment. Nematodes, tardigrades, rotifers, mites and tiny insects move in to feed on the living

Patterned ground from the air. The tundra shrinks in winter cold, cracking along planes of weakness. The cracks fill with ice and soil. Summer warming causes the tundra to expand, and pressure forces the edges of the cracks into ridges. When the tundra floods in summer, the ridges stand as banks surrounding shallow pools.

and dead plant materials (see below). Their activities help to keep minerals in circulation and increase the productivity of the developing soil.

So, good soils develop from poor ones. The process is painfully slow; time is an important constituent of every good soil. In the Arctic, soils develop especially slowly, for low temperatures slow down both plant growth and soil chemistry, and frost action produces new rock debris faster than plants can colonize it. Very dry tundra is too dry for soils to mature at all beyond the ahumic stage. Wet tundra is usually too wet. Permanently frozen ground (permafrost) below the soil keeps the water from draining away. Air cannot circulate, oxidation is slowed down and the soil becomes acid and sterile. The wet soil expands on freezing and contracts on thawing, shifting and churning within itself and never forming stable layers. Wet tundra develops characteristic polygonal and striped patterns of ridges and furrows, which change constantly and do not allow vegetation to become established.

The best Arctic soils, called *Arctic brown soils*, form on warm, well-drained tundra, usually close to the tree line or under a mat of vegetation; the organic cover adds humus to the soil and protects

Mosses colonizing rocky desert soil, Ellesmere Island. These communities contain insects, mites and other tiny animals.

Patterned ground. Some plants prefer the damp troughs along the cracks, and benefit from the shelter provided by the slight hollow.

Right below:
Calcium-rich soils support a grassy heath community and provide good summer grazing for herbivores. Cape Thompson, Alaska.

Arctic willows (dark) and moss (bright green) on a slope moistened by springs. West Greenland.

the surface from violent climatic action. In standing water Arctic brown soils grade into marshy pool and peat deposits. Northward from the forest edge, trees are replaced by tundra grasses, slow-growing heath plants and lichens, which add less organic material to the soil. Arctic Brown Soils give way to *polar desert soils*, which have a lower organic content and lower productivity. North of the 5°C. July isotherm vegetation becomes very sparse, and polar desert soils grade into true ahumic soils and gravel pavements.

Tundra plant communities

The tundra supports a surprising wealth of plant species, including algae, mosses, lichens, sedges, rushes, grasses, low, woody shrubs, stunted trees, and small, tufted, flowering herbs. Greenland alone has nearly 500 species of ferns and flowering plants, including ninety different kinds of grasses. Typically, tundra vegetation is a low, continuous mat only a few inches high, trimmed to uniform height by winds and herbivorous animals. In sheltered corners where soils accumulate and are damp throughout the summer, the vegetation grows rapidly to produce thickets and rich meadows. Elsewhere, growth is usually slow; miniature tundra willows a century old have stems no thicker than a thumb, and no taller than an Arctic hare. Greenland and southern Iceland produce good grazing for sheep and cattle, and their meadows yield an annual crop of hay for winter feeding. Perennial flowering plants form a carpet of vivid colours in spring, and brilliant berries enrich the autumn tundra, providing food

Warm, well-drained tundra develops rich brown soils, which support varied vegetation. The vegetation protects the soil, contributes to its development, and attracts grazing animals which, through feeding, return minerals to the soil.

for such unlikely picnic companions as ptarmigan, voles and grizzly bears.

Tundra vegetation forms recognizable communities, similar throughout the whole Arctic region. Where soils are richest and summer rainfall adequate, on sunny slopes inland from the sea, miniature forests of willow, birch and alder form *forest tundra copses*, dense enough to hide a bear, and tall enough to satisfy many tree-nesting species of birds. Warm slopes, protected by snow in winter, which thaw early in spring, produce a rich sward of *tundra grasslands*. Grasses predominate, but mosses and lichens grow freely among them. Tundra grasslands are favourite feeding grounds of musk-oxen and a haunt of ground-nesting birds, especially waders. They are richest on acid to neutral soils and are especially well developed in sandstone

A thick carpet of "reindeer moss" (*Cladonia* species), actually a lichen. This is one of the staple foods of reindeer and caribou.

A single plant of reindeer moss. Lichens are a mixture of fungus and algae. Few grow as large as these moss-like clusters.

Brilliant bearberries enrich the Arctic moorland-tundra: food for birds, large and small mammals—and man.

willow scrub

forest-tundra

reindeer moss

marsh and moss

ahumic soils

tundra grassland

moorland

lichen-covered rocks

permanent snow, ice cliffs

fellfield

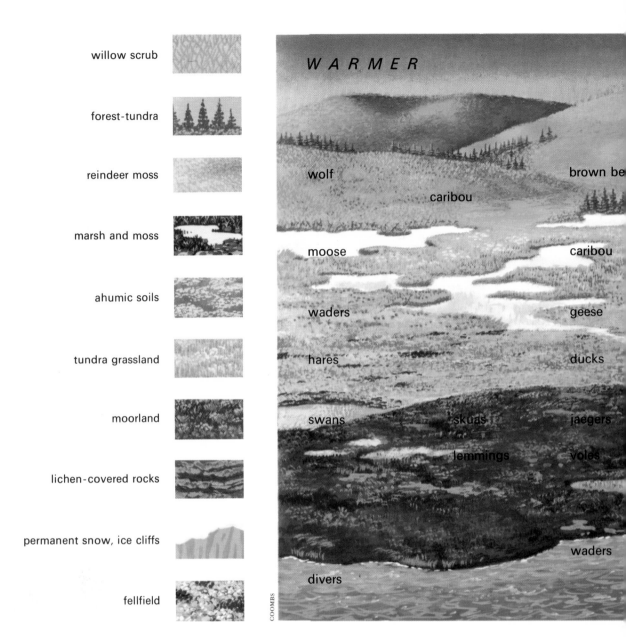

WARMER

wolf

caribou

brown be[ar]

moose

caribou

waders

geese

hares

ducks

swans

skuas

jaegers

lemmings

voles

waders

divers

COOMBS

The Arctic tundra. This imaginary landscape condenses into a single panorama a wide range of tundra conditions. Warmer climate and longer freedom from ice (left) allows the growth of forest-tundra. Cooler climate and newly-emerged ground (right) support only a meagre flora and fewer animals.

GREENE

Fellfield community of Arctic willow (*Salix arctica*) and *Polygonium viviperium*, with lichens and mosses. West Greenland.

musk oxen

yote   ptarmigan   Dall sheep

hmings   pikas

ctic fox   Arctic hare

musk oxen

polar bear

gulls and skuas

BREUMMER

Arctic tern nesting on the ground among the herbs and mosses of a tundra plant community.

Catkins emerge before leaves on the dwarf willows of the tundra.

Dune grasses. Great black-backed gull nesting on a sand dune stabilized by coarse grasses.

Lichens develop on stable surfaces. This community has formed on a cast caribou antler.

Well-established river bank community with tall willows growing close to the water. Northern Norway.

Midsummer in Iceland. Red-necked phalaropes hunt for emerging aquatic insects in a tundra meadow stream.

Pond-edge community in August, with cottongrass in flower. Shallow tarns and ponds form during the spring melt, disappearing later with summer evaporation. This is good breeding ground for migrant waders which feed on aquatic insects. West Spitsbergen.

regions. On gravelly, basic soils which dry out quickly in spring, grasses give way to sedges and reindeer moss (lichens of the genus *Cladonia*), and a thin growth of dwarf willows. This is the *Arctic steppe* community, sought out by reindeer and caribou during their summer wanderings on the tundra.

Herbslopes are well-drained slopes with good winter snow cover, which thaw slowly from early spring onward and have a long growing season. With good soils they develop a rich carpet of low flowering herbs including saxifrages, poppies, campions, shinleaf and Arctic bluebell. Peaty soils carry an *Arctic moorland* vegetation of heaths, heathers, bilberry, crowberry and many other small shrubs with strong, woody, wind-resistant stems and tiny leaves. *Fellfield* is the barest, driest moorland, too exposed to form peat and generally unprotected by snow in winter. Heath plants hug the ground, protected by boulders and patches of sedges, rushes and coarse grass. Vegetation is thin, and much of the gravelly soil is exposed. This is not good pasture for large grazing mammals, but Arctic hares, lemmings and bighorn sheep, which trim plants close to the soil, make good use of it.

Snow patches which remain for ten or eleven months of the year develop a characteristic *snow-patch community* of mosses and low shrubs. Nothing else will flourish in the short growing season which remains after the snow has gone. Dwarf willows thrust their shoots above the snow; the mosses wait, sometimes gaining benefit from the filtered light which penetrates the snow. Occasionally a tiny greenhouse forms immediately above the soil, giving the vegetation a brief spell of almost sub-

tropical warmth. If the snow persists in a cold summer, the plants simply wait until next year.

Fens, marshes and *bogs*, the wetland communities of the tundra, form where underlying permafrost holds melt-water close to the surface. Shallow, sluggish streams become choked with mosses and marsh plants, which gather clay and silt about their stems and form shifting islands in the stream beds. Shrub willows, sedges and rushes help to bind the soil, producing acres of water-sodden marshland and peat bog. Shallow lakes—some recent, others dating back to the last glacial period, and all frozen for three-quarters of the year—develop a

Herbslope with flowering saxifrage growing in a pocket of rich soil. Herbslopes provide good cover for small mammals and breeding birds.

Arctic lupins add nitrogen to the mineral-starved soils of the tundra, enriching it locally to the benefit of other plants.

distinctive lake-edge community of aquatic plants, including burr-reed, mare's tail and water buttercups, and dense mats of filamentous algae. *Salt-marshes* form close to the highest tide levels of the Arctic coast, with a characteristic flora of rushes, sea grasses, goose grass and sedge. *Sand dunes* carry scattered communities of lyme grass and succulent sea chickweed and oysterleaf, which help to bind and stabilize the sand.

Rock surfaces, both bedrock and stable scree slopes, have a distinctive and often highly coloured community of lichens. Some are appressed closely to the rock. Others form a thin covering of grassy texture, spreading over hundreds of acres in a dense, uniform carpet. Reindeer mosses grow freely on consolidated scree slopes, and other lichens flourish on any other dry, stable surface which is neither exfoliating (i.e. splitting due to frost action) nor suffering constant abrasion from wind-driven sand and snow. Mosses grow in the damper interstices between rocks.

Invertebrate animals

Though the bare Arctic tundra is seldom completely sterile, the protection and insulation of plant cover is needed before communities of macroscopic (i.e. visible) animals can form. Practically every kind of tundra vegetation has a fauna of small-to-tiny animals living among it; the more complex the plant community, the more stable and complex the animal community can become. Most of the animals are insects; mites, spiders and snails are also present. Most numerous are the herbivorous insects, a mixed bag including springtails (Collembolae), beetles, plant-sucking weevils, bugs and lice, craneflies, caterpillars of moths and butterflies, and the larvae of sawflies. These feed either directly on the living plants, or on debris

Three species of northern butterfly which reach the Arctic tundra. Duller colouring (left) is characteristic of tundra specimens. Top to bottom: *Colias palaeno, Clossiana selene, Erebia ligia.*

Arctic heather and Lapland rhododendron flourish on the acid soils of the Canadian Barren Grounds, attracting pollinating insects.

WARD

Dragonflies and damselflies have a larval development in fresh water. The cold lakes of the tundra do not favour their growth. Though common in the subarctic they are rare visitors to the tundra. Above: Dragonfly *Libellula quadrimaculata*. Right: damselfly *Enallagma cyathigerum*.

Spider web. The flying insects of the tundra provide plentiful summer food for spiders.

SIMPSON

BREUMMER

Mosquitoes emerge from lakes and ponds in summer. A nuisance to man, they are an important food of tundra birds.

Warble-fly larvae under caribou skin. These maggots would soon have emerged through the skin and pupated in the ground, emerging later as flies to lay their eggs on caribou fur.

BREUMMER

WARD

from plants, or even on the fungi which permeate decaying vegetation. Springtails are especially numerous. Normally well hidden among the moss stems and lichen bases, they emerge in millions on warm summer days, darkening the surface of snowbanks with their grey-blue, millimetre-long bodies. A square metre of poor, mossy ground supports two to three thousand springtails. The same area of rich Arctic brown soil with grasses and herbs might contain half a million or more.

Springtails and their fellow herbivores are eaten by carnivorous beetles, spiders and mites. Other insects prominent on the tundra include bumble bees and dipterous flies, which drone steadily through the summer. Blowflies, dung-beetles and burying beetles go busily about their task of returning nutrients to the soil. Butterflies and moths, usually drab versions of Eurasian and American species, flutter over the vegetation keeping low out of the wind. Parasitic ichneumonid wasps buzz among them, seeking caterpillars to lay their eggs in. On the wetter tundra lacewing flies, gnats and midges emerge constantly from fresh water streams and lakes, where their larvae have been growing. Dragonflies and damselflies are rare but colourful visitors from the Subarctic fringe. Very few Arctic localities have been examined closely to see what insects and other small animals are present, though there are certainly plenty. Greenland alone has over 600 species, including forty-six species of springtails, twenty-six butterflies and moths, and about the same number of beetles.

Two memorable Arctic insects are the mosquito and the blackfly. Mosquitoes emerge from lakes and ponds during June and July, forming huge swarms which rise like smoke above the tundra. Swarming mosquitoes are mostly males and relatively harmless; only the females bite, and they move singly or in small groups, searching for a meal of warm vertebrate blood which will enhance their egg production. The irritating buzzing and painful bites of Arctic mosquitoes keep man and other warm-blooded animals constantly on the hop through the early summer. As mosquitoes die back, blackflies take over. These are tiny swarming flies of the family Simuliidae, related to the sandflies of warmer regions. Emerging from water during two or three weeks of August, they buzz and bite furiously. Neither species penetrates to the far north; the coldest Arctic lands are mercifully free from biting flies.

Caribou and reindeer suffer greatly from mosquitoes but are also plagued by a pest of their own—the reindeer warble fly. This is a large, noisy fly which in summer lays its eggs among the warm abdominal fur. The larvae burrow into the skin and spend an agreeable autumn and winter browsing on their host's flesh. By spring nearly every adult reindeer and caribou contains several dozen, fat, contented maggots under the skin of their back and flanks. Emerging through the skin, the maggots fall into the ground, pupate, and turn into adult flies, which take wing in June and July to infest more reindeer and caribou.

Freshwater lakes and ponds of the tundra have a rich invertebrate fauna of their own. Like the sea, they develop a summer plankton of algae and crustaceans, but the larvae of many flying insects also feed in the bottom mud, among the weeds, or on the surface. The largest lakes are often the poorest, for they take a long time to warm through in spring, and may even keep a chilling core of un-melted ice through the summer. Small, shallow lakes warm quickly, developing a rapid turnover of energy and supporting a high density of zoo-plankton, insect larvae, snails and fishes.

Tundra herbivores

The largest grazing mammals which reside on the tundra are the musk-ox and the caribou. Musk-oxen are curious horned animals, now restricted to northern Greenland and the Canadian high Arctic tundra. With the humped profile and curved horns of a buffalo, and the woolly fleece of an over-endowed yak, they are isolated by zoologists in a genus of their own, somewhere between the sheep, goats and cattle. Bulls stand about one and a half metres high at the shoulder and weigh up to 700 pounds. Cows are slightly smaller. The shaggy brown coat has a matted underfelt fifteen centi-metres deep, windproof and snowproof, with long, shaggy, guard hairs trailing almost to the ground. The legs are short, the huge cloven feet slightly

The matted coat of the musk-ox helps it to winter on the bare northern plains of Ellesmere Island, Canadian archipelago.

Musk-ox herds form a defensive circle with horns out, young animals toward the centre.

asymmetrical. During the breeding season, which occurs in August, bulls emit a musky scent from facial glands.

The horns, thick at the base and tapering sharply, are plastered over the forehead. In ritual fighting during the breeding season bulls solemnly crash their horns together; the points, which could easily rip a rival to shreds, are reserved entirely for predators. Musk-oxen move in groups of ten to thirty, usually including a lead bull, several junior bulls, and cows with their calves. Attacked by wolves or polar bears, the herd forms a tight, defensive square with adults facing outward and calves in the middle. The ring of pointed horns is an unassailable fence, proof against every enemy but man. The herd forms a similar closed group in bad weather, reducing the heat losses of the calves substantially. In clear, calm weather a small cloud of water vapour, visible for miles across the tundra, forms over each herd of musk-oxen and helps to reduce their radiation losses.

Musk-oxen feed on high ground in winter, often in intensely cold areas where wind keeps the ground snow-free. Their forage—the frozen, dried remains of summer vegetation—is poor and unrewarding, but just adequate to keep these large animals alive through the winter. They live quietly, often huddling together for warmth, and losing weight steadily as they consume their reserves of fat. The late spring flush of vegetation restores them, and continuous, rich grazing through late summer and autumn replenishes their fat reserves. Calves are born in April and May and start to graze soon after birth, though they are also

Checkmate. Musk-oxen fighting for herd dominance on a chessboard of patterned ground. Food scarcity in their northern environment keeps the small herds moving over the tundra.

fed on milk for their first year or more. From May, adults start to shed their woolly underwear and grow new guard hairs to replace the old, worn, outer fur. During the transition they look like untidy knitting; it is sometimes difficult to remember that this unprepossessing beast has the secret of living through the coldest winters in the world's most northerly lands.

After the final (Würm) glaciation musk-oxen were widespread across northern Eurasia and America. They retreated northward with the ice sheets, and at the same time were hunted persistently by primitive man. By the end of the eighteenth century they were still present in a wide crescent across the southern Canadian tundra. Here they first met civilized man, whose muskets suddenly made nonsense of the stolid, defensive square which served them so well against other predators. Whalers, trappers and explorers, who wanted their meat and fur, marvelled at the stupidity of these great, ugly creatures, which stood four-square to be shot. By the early years of the present century, the wandering herds were reduced to a small remnant at the northern edge of their former range. Though many thousands of musk-oxen were butchered, climatic warming during the late nineteenth century may also have helped to drive them north, for they are well adapted to intense cold but seem unable to survive long spells of sleet, rain or strong sunshine.

Over the past half century the slaughter of musk-oxen has gradually ceased in Canadian territory and Greenland, and the species is now left to recover by itself in the far north. Breeding animals have been transferred to Svalbard, Nor-

The small, sturdy Peary caribou of northern Greenland and Ellesmere Island is well adapted to a meagre diet and a harsh environment.

way and Alaska in an attempt to rebuild stocks.

Caribou and reindeer are respectively the new-world and old-world representatives of a single species of tundra deer—*Rangifer tarandus*. There are many local geographical races varying in size, shape and colour. All are large, rather ungainly animals, superbly adapted for living in extreme cold. Caribou live in Greenland and northeastern Canada. Standing one and a half metres high at the shoulder, they weigh 500 to 600 pounds or more in full fat. Reindeer live in Greenland, northern Eurasia and islands of the eastern polar sea. Practically all are domesticated and in the care of herdsmen. They are smaller than caribou, standing one metre high and weighing only 200 to 250 pounds. Both caribou and reindeer cows are smaller than their bulls. They normally carry slender antlers keeping them through winter and early spring after the bulls have shed theirs. This is the only species of deer in which females wear antlers of any kind. It has been suggested that they use them in winter, to keep other cows away from feeding places in the snow.

Caribou and reindeer have coarse, compact hair, forming a dense covering up to five centimetres thick. Each hair is a hollow cylinder containing air cells; for its weight, caribou fur is probably the most windproof and best insulating fur known among mammals. The feet are flat and broad, spreading under the animal's weight and rebounding with audible clicks as it walks. The large antlers carry a broad, spade-like brow tine pointing forward over the face, which is said to protect the eyes from damage in dense woodland.

Caribou of the Canadian tundra ("barren

In summer the caribou of the Barren Grounds seek windy heights to keep themselves cool. Both sexes of caribou and reindeer have antlers.

Greenland caribou. The antlers, almost completely grown, are still covered with velvet, a living skin with large blood vessels which helps to keep the animals cool during the spring and summer. Later the velvet is shed and the antlers become weapons and social signals.

113

grounds caribou'') live in herds of up to 100 animals, massing together twice yearly for migrations which take them from summer to winter feeding grounds. Winters are spent in the shelter of the boreal forest north and west of Hudson Bay, where stock of smaller, non-migratory caribou ("woodland caribou") live all the year round. In the forest they feed on branches, grasses and reindeer moss, digging through the snow with their forefeet. From April they begin to mass and move northward, leaving the forest and spreading across the tundra. The calves, born in April and May, run with the migrating herds. On the tundra the animals feed voraciously, fattening after the hard fare of winter and laying down enormous reserves of food under their skin and in the body cavity. In summer, overheating becomes a problem, for they continue to add insulation as the weather warms. From April onward the antlers start to grow, providing a huge, additional surface of blood-warmed skin through which excess heat can be shed. On hot sunny days the animals climb to the ridges, seeking breezes and snow patches to cool them down. In July some move back into the forest for shade.

During late July and August the antlers harden

Domestic reindeer swimming to island pastures in northern Norway (part of the annual herding programme). Hollow hair, which helps to insulate them, also keeps them buoyant in the water.

Bull moose feeding on soft vegetation in a shallow forest-tundra pond. These are solitary, widely scattered herbivores which only occasionally leave forested areas for open tundra. The saucer-shaped antlers are shed and regrown annually.

Reindeer, the domesticated caribou of Eurasia. Caribou and reindeer vary locally in colour and size. Though many species and subspecies have been proposed, scientists now tend to group them all under the single species name *Rangifer tarandus*.

KINNE/COLEMAN

STONEHOUSE

Dall sheep feeding on last year's vegetation as it emerges from melting snowbanks.

Elk or wapiti, closely related to the European red deer, move out from the forest edge during the Arctic summer.

and lose their vascular surface, becoming the insignia and weapons by which the bulls establish a dominance hierarchy in the herds. Mating occurs in September, and the animals begin their southward migration toward the forest as winter closes in. Bulls lose their antlers in December and January. The caribou of Greenland, Ellesmere Island and other northern islands have no forest to migrate to, but they assemble in large herds in September and make local migrations to winter feeding grounds.

The vast herds of migrating caribou were estimated by early explorers to number several million animals. During the past two decades scientific survey has shown that the herds are reduced to between half and one million. Caribou suffered two forms of predation during their travels. The first, by wolves (p. 127), seems not to have affected their numbers or to have changed much over the years; wolf bands follow the herds, cull weak or disabled animals and a small proportion of the calves, but do not make serious inroads into the stock. The second important predator was primitive man. Both inland Eskimos and Indians of the Barren Grounds and forest lived successfully off the migrating herds for several centuries, in what seems to have been a balanced predation. The balance was upset in the nineteenth century, when the traditional skills of the primitive hunter were abandoned in favour of the gun; mild predation became mindless slaughter, followed predictably enough by a dearth of caribou at many points along the traditional migration route, and starvation for the tribes concerned. Protective legislation and research on the remaining stocks

have now been introduced, and there is some evidence that the caribou are no longer decreasing in number at a catastrophic rate.

Reindeer live a comparatively sheltered life. They migrate from north to south each year across the tundra in the care of their herdsmen; the movements are dictated by the annual cycle of seasons, so the animals are in some ways living as freely as their kin in North America. The reindeer provide meat, milk, cheese, skins and hides, bony implements, sinews, even windows (made from scraped intestines) for their masters. The herdsmen provide protection from predators and find good feeding grounds for their charges in return. Reindeer herding is a complete way of life, evolved over fifteen centuries and still practised by many thousands of nomadic people in northern Europe and Asia.

Largest of all living deer are the moose of North America and the closely related elk of Europe. Moose stand taller than a man at the shoulder, a large bull weighing over three-quarters of a ton. Moose and elk live mainly in the northern forests of Canada, Scandinavia, Russia and Siberia, feeding particularly on the soft vegetation of lake edges and wetter parts of the forest. In summer they occasionally visit the southern edge of the tundra, browsing on new vegetation in the tundra lakes. Bighorn sheep of Alaska and eastern Siberia, and Dall sheep of northwestern Canada, live in mountain tundra along the fringe of the Arctic. Like sheep the world over they graze close to the soil, growing fat in bare country which would seem incapable of supporting a rabbit. In winter they dig through the snow with their hooves.

OTT/COLEMAN

OTT/COLEMAN

Dall sheep carry a record of age in the growth rings of their horns. The larger one is about seven years old, the smaller about five.

Top: Dall sheep of north-western Canada and Alaska, a species of the mountain tundra. They vary in colour from white to grey or even black. Palest animals usually live farthest north. Cloven hooves grip the narrow mountain paths.

Small herbivores: cycles of abundance

The small herbivorous mammals of the tundra include hares, several species of voles and lemmings, muskrats, marmot and other rodents. Many of these are very small animals with a short life span (usually of one to three years) and an extraordinary capacity for reproducing rapidly when food is abundant. Tundra vegetation varies from year to year in quality and quantity, often in accord with climatic whims. In a poor year, when winter is hard and snow lies long on the ground, good food is scarce and the small rodents are only moderately fertile. Many adults and their offspring die from starvation, and the population remains low. In a good year when food is abundant, fertility is high and fecundity increases; voles and lemmings may produce four or five litters, each of five to six offspring, and many parents and offspring survive to breed in the following year. Two good seasons in succession may produce forty or fifty small animals where there were two before. After three good seasons the ground begins to seethe with them; voles and lemmings pop up everywhere, move uphill or downhill, eastward or westward in seemingly organized columns, and die in thousands crossing sea ice, lakes and mountain ranges. Then the food supply gives out, the population crashes, and only a very few animals are left to continue as before.

Fluctuations of this kind occur locally, usually over a few hundred square miles and not necessarily in synchrony with neighbouring areas. Herbivorous birds (e.g. ptarmigan—page 140) show similar fluctuations in numbers. The small species of birds and mammals fluctuate rapidly, at intervals of three to four years. Larger species have a longer cycle; tundra hares, for instance, fluctuate in cycles of ten to thirteen years. The situation is complicated by predators—snowy owls, jaegers, Arctic foxes and others—which respond to changes in prey density with cycles of their own. At the first sign of a local eruption predators move in from other regions where business is slow. Because food becomes plentiful their fecundity rises and increasing numbers begin to bear heavily on the stocks of prey. When the herbivores begin to overtax their food supply there is a brief bonanza for the predators, followed by starvation.

Arctic lemmings. Differing locally in colour and size, lemmings have a similar ecological role throughout the Arctic. Adults are between 4 and 6 inches long (excluding the stubby tail) with dense fur which almost hides their ears.

118

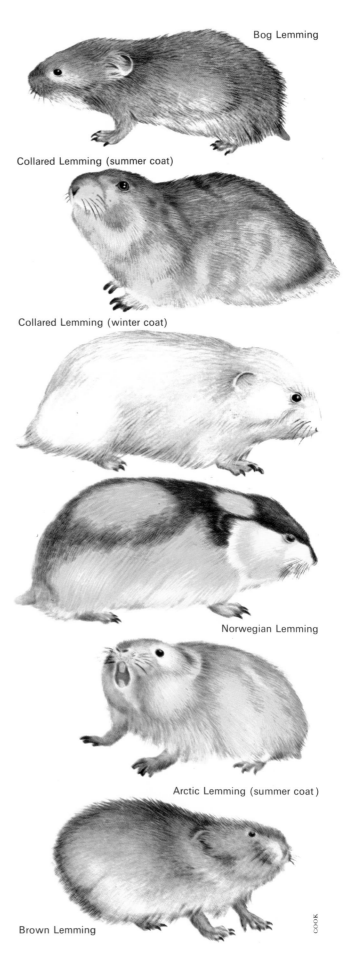

Bog Lemming

Collared Lemming (summer coat)

Collared Lemming (winter coat)

Norwegian Lemming

Arctic Lemming (summer coat)

Brown Lemming

COOK

Meadow Vole

Insular Vole

COOK

Tundra Redback Vole

OTT/COLEMAN

Pikas, closely related to rabbits, store hay in their scree-slope burrows to see them through the winter.

Arctic voles. Voles are smaller than lemmings, and tend to live close to the forest edge rather than an open tundra. Like lemmings, they are herbivorous, and remain active in burrows under the snow throughout winter.

Meadow vole in summer vegetation. Like most other small mammals of the tundra, meadow voles remain active under the snow in winter, eating insects in the soil, seeds and vegetation.

BARTLETT/COLEMAN

OTT/COLEMAN

Hoary marmots hibernate deep under scree slopes. Large enough to store body fat, they can last the winter without feeding.

Muskrats live in damp forest, building lodges of reeds, stick and other vegetation in marshy pools. They swim powerfully, using their tail (flattened from side to side) as a propeller. Dense, waterproof fur allows them to penetrate far to the north in the forest-tundra of Labrador and the Mackenzie River delta. The fur is highly valued by trappers.

OTT/COLEMAN

Some of the earliest information about these regular cycles of dearth and abundance was gained from nineteenth century fur records of the Hudson Bay Company, which showed clearly that many species fluctuated in numbers with more or less regularity. Later work gave a clearer picture of how cycles in predator species (lynx, fox, coyote, wolf and others) followed closely on cycles in snowshoe rabbits, their main prey.

Lemmings, perhaps best known of all for their four-year cycles of abundance, look like large, prosperous mice and weigh two to three ounces. Widespread across the Arctic, they form many local populations which have given rise to a confusing array of specific and subspecific names. Basically there are three kinds of lemming in the Arctic. The genus *Lemmus* includes Norway lemmings of Scandinavia, Siberian lemmings of northern Asia and the Siberian polar islands, and brown lemmings of Alaska and the Canadian Arctic. The genus *Dicrastonyx* includes the Arctic collared lemming of Asia, the Greenland collared lemming of Alaska, Canada and Greenland, and the Hudson Bay collared lemming of the Ungava Peninsula and Labrador. The genus *Synaptonyx* includes only one Arctic form, the northern bog lemming of Subarctic Canada which penetrates north in Ungava and Labrador.

Greenland collared lemmings are white in winter, grey in summer, with a reddish-brown band or collar across the back and flanks. Like most other species of lemming, they live underground in dry soil, digging well-organized systems of tunnels with sleeping chambers, latrines and other conveniences. They remain active in winter, living under the snow in an additional system of galleries and tunnels, well sheltered from the weather and in close contact with the buried vegetation. In years of abundant food they begin breeding early, producing the first of their four or five litters in March. During population explosions they leave their burrows and travel, usually in very large numbers. Other species have a very similar way of life. Similar again are the voles of the genera *Microtus* (meadow, insular and Alaska voles) and *Clethryonomys* (tundra redback vole), which reach the Arctic in America and Eurasia. These are mostly smaller than lemmings, weighing one to

Arctic ground squirrels hibernate in communal burrows for part of the winter, emerging on warm days to feed.

Snowshoe or varying hares are white in winter (right) and dark brown in summer (right, middle). The black tail and brownish underfur distinguish this species from the Arctic hare.

Porcupines are tree-living rodents, with a defensive covering of brittle-tipped quills. They visit the forest-tundra and bush-covered tundra in summer, in search of soft willow stems and other young vegetation.

OTT/COLEMAN

OTT/COLEMAN

MADDEN

Arctic hares are snow-white throughout the year in the far north. Southern stocks on the Canadian mainland turn brown in summer. Their fur is soft, but valued by Eskimos for its warmth.

two ounces, but their feeding, reproduction and population explosions follow similar patterns.

Arctic ground squirrels, widely distributed across the tundra between Hudson Bay and Alaska, and hoary marmots of western Canada and Alaska, are rather larger species which hibernate for five to six months each year. Ground squirrels make elaborate tunnel systems in which whole communities live and hibernate together. Hoary marmots usually live deep in cavities among the boulders of rock slides and screes. Pikas of western Canada, living in a similar environment, spend their summers collecting grass and drying it in the sun. Their annual hay crop is then stored under the rocks of the scree slopes, to serve them for food and bedding during their winter retreat. Muskrats, aquatic rodents weighing one to two pounds, and porcupines, weighing up to thirty pounds, are forest animals which occasionally find themselves foraging on the southern edge of the tundra.

Arctic hares are large, snow-white hares weighing nine to twelve pounds. In the high Arctic of northern Greenland and Ellesmere Island they are pure white (except for black ear tips) all the year round. Further south, on the Canadian mainland, they turn slightly grey or brown in summer. The tundra hare of Alaska is closely related to the Arctic hare, if not identical with it. Snowshoe or varying hares are widespread in the forests of Subarctic and temperate North America, and seen occasionally as visitors to the tundra. They are smaller, rabbit-sized hares weighing only three to four pounds; white in winter, they turn a rich brown in summer. Blue hares of Eurasia are similar, though slightly larger, and occupy a similar niche in the old world.

True Arctic hares live in loose groups of up to 100 animals. They wander freely over the tundra throughout the year, often seeking high ground in winter to avoid deep snow. Small enough to fit neatly among the rocks and hide under vegetation, they can readily find shelter from the coldest winds. Mobile enough to travel far with little effort, they can practically always find enough deep-frozen vegetation to keep them going through the depths of winter. Against a snow background they are practically invisible, and often overlooked by hunters and predators. They breed from late April onward, usually producing a single litter of six or seven grey leverets.

123

Tundra carnivores

While polar bears range the Arctic sea ice in search of seals, three kinds of brown bears wander over Arctic lands in search of honey, berries and lemmings. Largest of the land bears is the Kodiak or brown bear of southern Alaska, which stands one and a half metres at the shoulder and weighs up to 1,500 pounds. It is only marginally an Arctic species. Grizzly bears of northwestern Canada and Alaska are next in size, standing one metre high at the shoulder and weighing up to 1,000 pounds. Smallest are black bears of North America, Europe and eastern Siberia, an extremely variable species, which usually stands less than one metre and weighs less than 500 pounds. In spite of a reputation for fierceness, the brown bears are normally mild and tend to be vegetarian; they could as well have been included among the large herbivores. Males are solitary and relatively sedentary, occupying large ranges which they patrol systematically in search of food. Females amble more widely, often with two or three cubs roistering behind. They feed on shoots, roots, leaves, grass, berries and wild honey; as opportunity offers they also take small rodents, eggs, insects and grubs, fish (which they scoop with skill from the rivers during breeding runs) and offal. Bears have no false pride, taking readily to garbage-collecting in state parks, and standing with paws outstretched for welfare handouts. In the far north they patrol the tundra through the summer and sleep through the winter, emerging periodically to search for a meal before continuing their sleep.

Brown or Kodiak bears, mainly subarctic, wander on the Arctic tundra in summer. Here they fish for salmon in an Alaskan river. This species is the largest of all living bears.

Grizzly bear of Alaska and north-western Canada. Varying in colour from yellow-brown to black, grizzlies usually have long, grey or white-tipped guard hairs, which give them their greyish or grizzled appearance.

GILLSATER

ROBINSON/COLEMAN

Though normally herbivorous, black bears never miss the opportunity of a good protein meal, usually scavenging after other animals have made the kill.

Dark form of the grey or timber wolf, Alaska. Wolves are the major predators of the tundra; they follow caribou herds culling sick and ageing animals. Man's vendetta against wolves destroys the balanced ecology of the tundra.

OTT/COLEMAN

The grey or timber wolf is a more serious-minded predator, ranging widely across North America, Greenland and Eurasia. It reaches all the Canadian archipelago islands and the southern island of Novaya Zemlya, but has not penetrated further north in the Siberian or European Arctic islands. In the high Arctic many wolves are pale grey or brown; further south most are black, dark brown or dark grey, with pale forms in a minority. Wolves are long-legged, Alsatian-like animals, weighing up to 120 pounds. They hunt in mobs—basically family groups—of up to twenty animals. On the tundra, litters of five to ten pups are born in May. The pups stay with the family group for over a year, learning the tricks of social hunting on which the success of the species is based. In summer wolves feed mainly on small mammals and birds, eggs and carrion, living well in years when voles and lemmings are plentiful. Throughout the year they watch the movements of large mammals, especially caribou and moose; some wolf packs move with the migrating caribou herds, always ready to attack stragglers. Though ill-matched against healthy adults, they are often successful in dragging down calves or old and diseased animals. Wolves and men have never reached an understanding; even the tundra is not big enough for both of them. Hated and envied for his success as a predator, the grey wolf is shot on sight wherever he goes.

Coyotes are small dog-like carnivores of prairie and scrubland, weighing up to fifty pounds. Widespread across North America, they penetrate north to the Alaskan tundra. Coyotes are nimble enough to catch small mammals and birds in the

Tan-coated timber wolf eyeing an Arctic owl. Wolves of the high Arctic are pale, sometimes almost white.

Young coyote curled up in its den. Dens are usually natural hollows enlarged by the coyotes for their own use.

Following page: Coyote, the light-weight wolf of the tundra and prairie. Whereas wolves are socially organised, coyotes run singly or in loose packs, confining their hunting to small game.

Arctic foxes have the thickest fur, for their size, of any polar animal. Those of the high Arctic turn creamy-white in winter (right); southern stocks, like the Bear Island Arctic fox (above), remain grey or blue throughout the year. Their prey includes small mammals and birds, which they store to feed on during the winter.

undergrowth; lacking the weight and social organization of wolves, they cannot usually attack the larger mammals with any hope of success. They live in underground dens or burrows, producing a single litter of five to eight pups annually in April or May.

Arctic foxes are the smallest of the dog-like predators, seldom weighing more than fifteen pounds. They are completely circumpolar, living on every tundra coast and island and often wandering far out across the polar sea ice. Slightly smaller than red foxes, they have small, round ears and a very bushy tail. In summer they have a thin, grey-brown coat, which matches the tundra well. In winter, those of Canada and the far north turn rich creamy white, with a dense and beautiful fur much prized by furriers and their customers. Southern Greenland and Eurasian stocks tend to turn blue-grey in winter. Arctic foxes produce one or two litters each year, in May and occasionally again in August. Normally, there are six to eight cubs per litter, but a dozen or more may be born in years when lemmings or hares are abundant. Not surprisingly, numbers of this species fluctuate from year to year.

Summer food includes small mammals and breeding birds with their eggs and young. Arctic foxes are especially attracted to cliff colonies of sea birds, where they snap up any gull or auk unwise enough to nest within their reach. They store food in larders, usually in rock crevices beyond the reach of larger predators, returning to them in winter when other supplies fail. They are reported to follow polar bears across the sea ice in winter, nibbling scraps but keeping well out of reach when their host makes a kill. Red foxes are often seen on the American and Eurasian tundra in summer, emerging from the forest to feed on the abundance of birds and small mammals.

The mammal family Mustelidae, which includes weasels, otters and other predatory carnivores, is well represented in the Arctic. Largest of the terrestrial forms is the wolverine or glutton, a dark brown, bear-like creature, compact and powerful, weighing up to sixty pounds. Wolverine live in the forest and forest-tundra of Canada and

Wolverine; small, powerful predator of the forest and tundra. They are reported to attack caribou and other large animals, and though their reputation for ferocity may be exaggerated, they are remarkably strong for their size.

Red foxes of the forest penetrate far north across the tundra in summer to feed on smaller mammals and birds. This animal, still in winter coat, sleeps comfortably in the snow.

Eurasia, where they have gained an extraordinary reputation for brute strength and ferocity. Trappers swear that wolverine patrol their traplines, cunningly taking bait and stealing snared animals. They feed on carcases killed by other predators, munching the bones and skin which wolves and bears cannot deal with, and often take live prey on their own account. Wolverine breed in May, producing a litter of two or three cubs in shallow, snow-drift dens. The cubs travel with their mother, taking over a year to reach independence. The fur is water-repellent and does not hold ice, and is much in demand as a trimming for parka hoods.

The otters are large, lively aquatic mustelids with dense, waterproof fur, which are well enough insulated to penetrate far above the Arctic boundary. Weighing up to thirty pounds in Eurasia, slightly less in America, they are usually found in rivers and lakes, feeding on fishes, rodents, birds and any other small living prey. Otters wander widely in small family groups, visiting the tundra and northern estuaries in their travels but never straying far from water.

Mink, which are very much smaller mustelids, semi-aquatic with webbed feet and dense, dark brown fur, are widespread in North America and Eurasia. The North American mink extends northward to the tundra of Alaska, northwestern and northeastern Canada. Weighing up to two and a half pounds, they feed both on land and in the tundra streams and lakes, taking small mammals and birds. The marten of North America and pine marten of Eurasia are mainly forest animals, similar in size to mink but mainly terrestrial. They too make a living from birds and small mammals, invading the tundra edge when food is plentiful.

In its luxurious fur coat, the mink lives close to water in the forest and forest-tundra of Eurasia and North America.

Common otter, an aquatic animal which occasionally visits the tundra lakes and rivers. Otters feed on fish and other aquatic foods, including insect larvae and molluscs.

The pine martens of Europe and Asia are similar in ecology and behaviour to the martens of the new world, though slightly heavier. They are seldom found far from trees, in which most of their life is spent.

Short-tailed weasel or ermine in winter dress and (right) summer dress. This animal is closely akin to the stoat of Europe and Asia; the year-round black tufted tail distinguishes it from the American least weasel and Eurasian weasel. Larger than the weasel, it preys on hares and smaller rodents.

Dark-furred martens are forest-living carnivores of North America, related to weasels, otters and wolverine. They reach the Arctic at the northern forest edge, living on small mammals, insects, birds, and other creatures caught on the ground or in trees.

Right, below: The European weasel is small enough to make a living from mice and voles.

Weasels and stoats are much smaller mustelids, well represented in the far north. The short-tailed weasel or ermine of North America is almost identical with the stoat of Eurasia, weighing between four ounces and one pound. Slinky and supple, fierce and alert, they weave among the rocks and vegetation of the tundra in search of prey. In summer they are reddish brown above and creamy white underneath. In winter, northern stocks turn entirely white, except for the tip of the tail which remains black throughout the year. Voles, lemmings and birds are probably staple foods, though hares and rabbits are not too large to escape their ferocious attentions. A single litter of four to eight young is produced annually in April or May. The weasel of Eurasia and least weasel of North America are similar but smaller animals, weighing respectively about four and two ounces. They are tiny enough to follow voles and lemmings into their burrows, kill and eat the occupants, and take over the nests for their own breeding. Like ermine and stoats, weasels of the far north turn white in winter. They breed once or twice each season, producing litters of four to five offspring.

Shrews are the tiniest four-legged predators of

Red Bat

Hoary Bat

COOK

Bats feed on flying insects; these two species are summer visitors to the Arctic, returning south in autumn when the brief season of insect abundance is over. They are seldom found far from forest-tundra regions.

Arctic shrews. Like tiny mice, shrews live close to the earth and burrow in soil and vegetation. They are entirely carnivorous, eating insect larvae, beetles, worms and other soil animals. With an enormous surface area for their size, their heat losses are great; they eat perpetually to keep warm, and seldom emerge into the open.

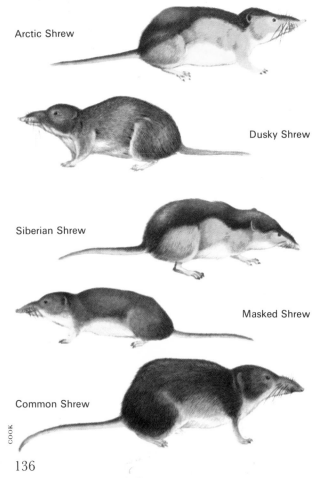

Arctic Shrew

Dusky Shrew

Siberian Shrew

Masked Shrew

Common Shrew

COOK

the tundra, generally weighing less than half an ounce. They feed almost exclusively on insects and other invertebrates, which they snuffle and dig from the ground with a long, mobile snout. Masked, Arctic and dusky shrews penetrate far into the northlands of North America. The common shrew of Europe and Siberian shrew of western Siberia are also tundra species. The Pribilov and Aleutian Islands have local populations of shrews. Tundra insectivores also include two species of bats; the red bat and hoary bat of North America make annual migrations to the north of Canada, reaching the Arctic in the Hudson Bay region.

Only two very closely related forms of cat have invaded the northlands—the Canadian lynx of North America and its Eurasian equivalent (*Lynx lynx*). The Eurasian animal is a squat, powerful creature weighing up to sixty pounds; the Canadian lynx seldom exceeds half that weight. Both are pale brown, with black tufted ears and sideboard whiskers, and a short stubby tail with black tip. They are mainly forest animals, adept at tree climbing and leaping. In Labrador, Alaska and northern Europe they emerge from the forest in winter; their main food is hares, with lemmings and voles as alternatives when hares are scarce. Lynx live in dens deep among rocks or under thick vegetation. They produce two or three kittens in April, and the kittens undergo several months' apprenticeship with their mother before being left to hunt on their own.

The Canadian lynx, a bob-tailed wild-cat, has a beard and tufted ears. These are forest predators which often spring from trees to catch their prey. They are winter visitors to the tundra.

BREUMMER

Lapland buntings move south for the winter. They arrive early on the tundra in spring, when their breeding plumage matches the background of dead vegetation in which they nest.

Snow buntings winter far north of the Arctic Circle, feeding on grass seed in the snow-free regions of the tundra. In summer they arrive early on the extreme northerly shores of the Arctic.

Birds of the tundra

With its rich vegetation, streams, lakes and wealth of insect life, the summer tundra offers a variety of niches or ecological slots for animals to occupy. Only a few are taken up by mammals. Though many more kinds of mammal could make an excellent living from the summer tundra, we must suppose that only the select few can cope with the hardships of the same environment in winter. The vacant summer niches are filled, logically enough, by migrant animals. Some are mammals from the forest edge. Most are birds—swans, geese, ducks, waders, predatory birds, songbirds—from the temperate zone, whose mobility gives them an advantage over mammals. About 100 species of birds nest on the tundra in summer; all but half a

OTT/COLEMAN

dozen hardy species leave for warmer climates in autumn.

The overwintering species include snow buntings, which breed in northern Greenland and on all the islands of the polar sea, the tundra coasts, and in some temperate regions, including Scotland and southern Scandinavia. In winter they move south from the extreme northern limits of their breeding range, but generally remain within the Arctic region. Their main food is grass seed, culled from dead stalks or picked laboriously from gravel and snow drifts. They also feed like sparrows about human settlements, especially in winter. April sees them moving northward across the snowfields and pack ice to the breeding grounds; by May they begin to stake out territories and sing. In summer they feed and raise their families on insects.

Lapland buntings breed further south on the tundra and winter in the temperate zone. Hornemann's redpoll, the Arctic representative of the common temperate redpoll, also breeds on the tundra coasts of Eurasia, Greenland and North America, and tends to remain sedentary in winter. Often found among the stunted trees of the forest-tundra, it feeds on seeds of deciduous trees as well as grasses. Common redpolls reach the tundra edge and in places may overlap and interbreed with Hornemann's redpoll.

Rock ptarmigan, like snow buntings, breed in the far north of Greenland, Ellesmere Island,

Redpolls breed far north in the subarctic and Arctic. A high Arctic form, Hornemann's redpoll, similar but with paler plumage, is one of the northernmost breeding birds.

Rock ptarmigan (left) and willow ptarmigan are speckled brown in summer and white in winter. Males have prominent red combs or "eyebrows", females slightly smaller ones. These birds feed on the ground, digging in the soil with feathered feet to find seeds, shoots and insects.

Male willow ptarmigan (right) during the spring change of plumage, female (below) in summer plumage.

Bewick's swan (below) of Eurasia and whistling swans (above) of North America head north for the Arctic in spring to breed on tundra lakes. They feed on lake weeds pulled from the bottom in shallow water.

Snowy owl chick shedding the last of its grey nestling down. Snowy owls feed on small mammals; their numbers fluctuate from year to year as mouse and lemming populations vary.

Tundra scavengers. Coyotes, ravens and a magpie ensure that even in winter nothing edible is wasted in tundra economy.

throughout the Canadian archipelago and on all the tundra coasts of the great continents. They too remain on the breeding grounds throughout winter, shifting away from areas of intense cold or heavy snow. In very cold weather they burrow into snow-drifts for warmth, and in search of the shoots, leaves and berries on which they feed. Closely related willow ptarmigan breed on the mainland tundra and the southern islands of the Canadian archipelago, but do not penetrate so far north as rock ptarmigan. Both species are speckled brown in summer and white in winter; the tail feathers remain black throughout the year.

Ravens of the familiar European and American species breed far north into the tundra, on Wrangel Island, throughout eastern Siberia and Alaska, the Canadian archipelago, Greenland and Iceland. Largest and most versatile of the crows, they feed on every kind of organic material from berries to dead fish, from stranded plankton to the afterbirth of seals. Ravens winter close to their breeding areas, croaking happily through the coldest weather and surviving when all else. is dead. Ever receptive, like shiny black trashcans, they play a useful role in Arctic economy. Snowy owls, most characteristic of the sedentary, high Arctic species, breed and

Snow goose; pure white except for the wingtips, the North American species winters on warm coasts of the United States and flies north each year to the tundra to breed. Wanderers occasionally visit Iceland and Europe.

winter in northern Greenland, North America, the Canadian islands, across northern Eurasia and on Novaya Zemlya and Wrangel Island. Most prominent in lemming country, their movements are determined by their prey. Where lemmings, voles or hares begin to increase locally, there the owls gather. In winter they hunt silently over the snow for ptarmigan and smaller birds.

Summer visitors to the tundra include both aquatic and terrestrial birds. Bewick's swans of Eurasia and whistling swans of America head north in April and May to nest in marshy, low-lying ground among shallow tundra pools and waterways. Their food is floating and submerged vegetation, which the long neck helps them to reach. A host of dabbling ducks, including mallard, wigeon, gadwall, teal, green-winged and blue-winged teal and pintail, share the fresh-water environment. Together with the diving ducks (page 92), the divers or loons (page 91) and the phalaropes (page 76), they invade the Arctic in huge flocks as soon as the thaw begins and exploit the fresh water habitat fully during the long hours of summer daylight.

Most spectacular of summer migrants are the geese, which fly northward each summer in

Arctic geese. Geese are grazing birds, completely at home in wet grasslands. The Arctic in spring suits them well, and thousands fly north each year across North America and Eurasia to rear their families on the lush new growth of the tundra.

Pink-footed Goose

Blue Goose

Ross's Goose

Brent (Brant) Goose

Barnacle Goose

Red-breasted Goose

White-fronted Goose

Bean Goose

Mallard

Pintail

Gadwall

Widgeon

Teal

Bluewing Teal

Greenwing Teal

These ducks visit the Arctic to breed on the tundra wetlands, dabbling for insect larvae in shallow waters.

Oystercatchers use their slab-sided bill for opening cockles, prising molluscs from their shells, and digging in soil.

Golden plovers breed in the Arctic, wintering in southern Asia and the U.S.A. A comparable species breeds in the European Arctic and subarctic, wintering in southern Europe.

The long sickle bill of the curlew sweeps and probes equally effectively in long grass or shallow water.

garrulous, honking skeins to feed and raise their families on the tundra. Brent geese breed along the cold tundra coasts of Canada, Greenland and Eurasia, including the Canadian and most of the Siberian polar islands. They nest on estuary shores, islets and wet coastal plains, never far from the sea, and feed on fresh grass and new shoots of the tundra herbs. Barnacle geese of Greenland, Iceland, Spitzbergen and Novaya Zemlya and red-breasted geese of the eastern tundra breed in more rugged country, often nesting on cliffs and steep rocky outcrops to avoid ground predators. White-fronted geese, which breed throughout the

Arctic, and the several forms of bean and pink-footed geese of the Eurasian tundra and Iceland nest in low-lying marshy places. Ross's geese and snow and blue geese occupy a similar niche in the Canadian and Alaskan Arctic. These too are grazing geese, feeding on grasses and other low vegetation in summer, and wintering mainly on coastal flats, estuaries and wetlands of the temperate zone.

Over thirty species of waders or shore birds migrate annually to the tundra. Nearly all are familiar species of the European or North American winter shore. Some, like the grey plover and turnstone, are well known on southern hemisphere shores as well. The waders are, without exception, carnivorous. In summer their food is mainly insects and insect larvae, small fishes and molluscs, and crustaceans. A few also take seeds, roots and new, green shoots as additions to their insect diet. The many different shapes of bill indicate the range of ecological niches occupied by the waders. Oystercatchers, with their enormous, coral-red, probing bill, reach the Arctic only in Iceland and northern Europe. The plovers and turnstones, with their short stout bill, are well represented in the far north. Grey or black-bellied plovers, European and American golden plovers, semi-

Sand martin. A cliff-nesting species, it catches insects on the wing.

DALTON/NHPA

Smallest of all the waders, the little stint breeds in damp, marshy scrub and winters on warm coasts and estuaries.

HAKANSSON/COLEMAN

Black-tailed godwit, a tundra nesting bird. The warm breeding dress colours are replaced by drab winter plumage on migration.

PORTER/COLEMAN

COLEMAN

Greenland wheatears breed in Arctic Canada, Alaska and Greenland. The short, stout bill is a general-purpose probe, suited to a wide variety of foods. Wheatears are among the earliest birds to arrive on the tundra each spring.

Meadow pipit with nestlings in a well camouflaged grassland nest. The closely related and similar red-throated pipit spreads far across the Arctic tundra.

PARK/COLEMAN

Great grey owl, a woodland species that occasionally spreads to the tundra in Canada and Scandinavia. Its main food is small animals, on which it swoops in the half-light of the evening.

Right: Redwing. A small thrush-like bird familiar as a migrant in Britain, it breeds in Iceland and northern Europe.

Right, below: Familiar in Europe, the yellowhammer penetrates north to breed on the tundra.

MIDDLETON/COLEMAN

palmated and ringed plovers, dotterel and turnstones all have wide or restricted ranges on the tundra. They breed between May and August in well-camouflaged nests among the low vegetation. Whimbrel, curlew, and Hudsonian, bar-tailed and black-tailed godwits sweep, probe and dabble with their slender, curved bills. Over a dozen species of slim-billed waders—sandpipers, stints and sanderlings—find niches on the northern coasts or islands. Furthest north in breeding distribution are the purple sandpiper, knot, dunlin and sanderling, which breed in Greenland and on islands of the polar sea. Other species including dowitcher, ruff, and Temmink's and little stints are more at home on the continental tundra.

Practically all of the smaller migrants are insectivorous. Rock, meadow and red-throated pipits, and white and citrine wagtails take insects from ground and vegetation; horned larks and sand martins catch them in mid-air. The small passerines, including wheatear, bluethroat, redwing, fieldfare and yellowhammer, take a mixed diet of ground-living insects, grubs and snails, seeds and berries.

Large raptorial or predatory birds include the resident snowy owl (p. 143) and the short-eared owl, a smaller species of world-wide distribution, which breeds thinly along the Arctic mainland shores. Great grey owls, mostly a woodland species, also move north to the tundra in western Canada. Both short-eared and great grey owls feed mainly on voles and lemmings. Gyr falcons and peregrine falcons breed on the mainland Arctic coasts, Canadian archipelago and Greenland coasts in summer, part at least of the breeding population remaining in the Arctic throughout the winter. Willow and rock ptarmigan, and small birds and mammals, are among their favourite foods. Merlins migrate north to the Mackenzie river delta, Labrador, Iceland and the European tundra, feeding mainly on small birds which they catch on the wing.

Following pages:
Man in the Arctic. Bloody trails converge on sealing ships in the pack ice. Eastern Canada.

Young peregrine falcon in cliff nest. Like gyr falcons, peregrines hunt live food, scanning the ground from high in the air.

A large falcon of the Arctic, the gyr falcon varies in colour from pale grey or white to dark rusty black. Greenland and other high Arctic birds tend to be palest. Gyr falcons take their prey, usually small birds, on the wing, or swoop to pick up small animals from the ground.

SIMPSON

Pola

Challenge: Man in the Arctic

Man evolved as a nomadic, socially-minded hunter of warm forest and grasslands. From a subtropical nursery in southern Asia he spread north, east and west, reaching south Europe half a million years ago and the north European plains some time later. Like many other mammals he lived close to the edge of the ice sheets during the glacial epochs, moving northwards across Eurasia as the ice retreated. He probably reached the Arctic coast during a warm interglacial or interstadial, when forest and grassland extended far beyond their present limits. The first human wave crossed Beringia to North America some thirty thousand years ago, moving southward before the final advance of the Wisconsin ice sheets (p. 47).

With his tender skin, naked body and lean, spindly limbs, man is not formed for polar living. Yet he has survived for more than twenty thousand years as a subpolar or polar species, sharing the coldest periglacial environments with shaggy musk-oxen, polar bears and caribou—all of which are far better equipped to deal with cold. Physical limitations of man include his lack of built-in weapons; from the start he has been forced to live by his wits. His wits have proved more than adequate for polar living. Sharper than teeth and more penetrating than horns, they have helped him to fatten on the best meat, and clothe himself in the world's most luxurious furs.

Native populations

The earliest human settlers of the far northern coast were inland folk who probably followed the great rivers northward. Hunters, trappers and river fishermen, they would have had no tradition of maritime adventure. They probably lived on wide wooded plains beyond the present tundra coast, in places which rising sea level has since made part of the sea bed. Earliest of all were probably the Yenesi-Ostyaks of western Siberia and the Yukaghirs and Chukchi of eastern Siberia, who are linked by languages and physical similarities, and seem to have spread northwards from a centre in the Siberian heartland. Early inhabitants of northern Scandinavia and the White Sea region were among the first settlers of Arctic Europe. North American Indians, who preceded Eskimos in the New World by many thousands of years, may also have spread into the far north.

Later waves of settlers in Eurasia included peoples whose languages are linked with those of Turkey and Mongolia—the Finns of northern

Eskimos hunted walrus annually on this Hudson Bay slaughtering ground. Though still plentiful, walrus range less widely than before, mainly because of man's heavy depredation.

Chukchi of eastern Siberia in traditional skin clothing. Eskimos of Alaska and Canada may be descended from these eastern hunters.

Scandinavia, the Samoyeds of north Russia, and the Tungus and Yakuts of eastern Siberia. These were nomadic like their predecessors, hunters first and then herdsmen, who became increasingly dependent on the great herds of reindeer and followed them northward. Their descendants continue the herding tradition, moving annually with the reindeer between tundra and forest.

The Chukchis of eastern Siberia are the coastal folk from whom Aleuts and Eskimos of North America are believed to be descended. Aleuts and Eskimos crossed the Bering Straits in boats some six to ten thousand years ago. Mariners and sea fishermen, they show clearly their Mongoloid origins in facial and other physical features. Aleuts restricted themselves to the Bering Sea region. Eskimos spread eastward along the coast of North America to the Canadian Archipelago and Greenland. Their descendants continue to live in small, semi-permanent settlements, usually within sight of the sea. In winter they hunt over the inshore ice, taking seals, walruses and bears. In summer they hunt and fish from kayaks or skin canoes, which carry them swiftly between the ice floes and are light enough for one man to lift from the water. Eskimos also hunt inland in summer, scouring the

Right: Canadian Eskimos are allowed to kill a limited number of polar bears for their own use. This bear, killed near Ellesmere Island, will yield meat, fat and bedding for an Eskimo family.

Far right:
Eskimo hunters in the far north take local seals for meat, oil and clothing. This non-commercial hunting is difficult, requiring traditional techniques of stalking and harpooning from kayak or ice, and does not endanger the stocks permanently.

Eskimos fishing with wooden harpoons at a stone weir. The fish, char or salmon, are dried and stored for winter.

tundra for caribou, musk-oxen, hares, foxes and smaller game. During spawning runs they fish for salmon in the tundra streams, using weirs and traps to increase the annual catch. In autumn they collect sea birds from cliff colonies. Travelling widely by dog sledge and boat, they live precariously from season to season and year to year. Centres of Eskimo population shift over the years as sections of the coast are hunted out.

Some Eskimos of the Hudson Bay and Keewatin regions moved inland and came to depend, like their Indian neighbours, on river fishing, trapping, and the annually migrating caribou herds. Greenland Eskimos never developed the agricultural possibilities of their country; though they hunted inland, they remained a sea-going community, with a nomadic habit which carried them to every point of the coast. Settlers from Europe became the farmers of Greenland.

The primitive peoples of the north evolved similar cultures and ways of life. There was little communication between those of the American and Eurasian Arctic, but the same stringent climatic conditions and limitations of materials forced similar cultural patterns upon them. They dress in similar style, using the fur and feathers available through hunting. Basic clothing includes breeches of stitched caribou, reindeer or bear skin, with a loose-fitting parka or coat of the same material. This combination is warm when the wearer rides his sledge, and sufficiently ventilated during activity to keep him cool. The hood is trimmed with wolverine fur, which does not freeze up when breath condenses on it. Other furs are used decoratively, partly to show the hunting

skills of the owner, but also in complete harmony with a sense of artistry which many northmen seem to possess.

Boots for use inland are made from leg skins of reindeer. On the coast and sea ice they are waterproof sealskin, stitched with a sinew which swells on wetting to close off the needle holes. Lined with socks of soft fur and packed with grass, sealskin boots freeze to the shape of the wearer's feet and remain comfortable all day. Big fur mitts complete the outfit. In winter Eskimos wear an undershirt of soft fur or bird skins, and sleep in rugs of polar bear or caribou fur. In very cold weather an additional layer of outer clothing keep them warm during long hunting vigils on the sea ice. Trade has brought new fabrics to the Arctic, and many northern folk now prefer mail-order catalogue parkas and gloves for everyday wear.

In warmer parts of the tundra reindeer hides, stretched over driftwood or birch poles, provide year-round tenting. Extra skins are added in winter for a lining, or a turf wall is built outside to keep the wind off. While travelling, Eskimos of the far north build their traditional domed houses of snow blocks. Safer in strong winds, they are also warmer and better suited to a wandering life. Snow igloos are kept warm by tiny lamps of seal oil, and by the cheerful presence of the Eskimos themselves.

Snowshoes and skis are used by reindeer herdsmen in the deep snow of the forest-tundra. Further north the snow is thinner and harder, and coastal Eskimos have little use for them. Sledges are the universal vehicles, in a variety of different designs to suit the raw materials available. Throughout the

Arctic dogs are used as draught animals, running in teams or family groups; locally they become pack-carriers, and help in rounding up reindeer or hunting. Sledge dogs vary in breed, size and character, from the lank, wolf-like malmute of northwestern Canada to the stocky husky of Labrador and Greenland and the chunky Samoyed of Siberia. Carnivorous, and able to fend for themselves in summer, dogs are most practical for coastal and tundra travellers. They are strong, robust, and splendidly protected in their warm fur against all but the very worst weather. Reindeer are the pack animals and sledge haulers of the Eurasian forest-tundra, and the cultural focus of those who herd them.

During his centuries of life in the Arctic, primitive man developed skills which brought him to terms with the environment. Though harsh, the terms gave him a living and a non-destructive role in Arctic ecology. Stone and animal products were his two raw materials. His populations were small, culled ruthlessly by cold and hunger, and the threat of starvation and promise of a better living elsewhere kept him nomadic. Without metals or sophisticated weapons he could seldom take more at a time than he needed from the environment, and lacked power to destroy it wantonly.

Eskimo in department-store parka, building a traditional snow house.

Sledge dogs at feeding time: Ellesmere Island. Though motor sledges replace huskies for local travel, the dog's ability to live on fresh or dried meat makes it invaluable for long journeys over sea ice and coastal plains.

Reindeer gather about a herdsman's tent in Norwegian Lapland. The herds and the folk who tend them move constantly through the summer in search of new pastures.

Norwegian Lapp in birch forest-tundra with pack reindeer.

Day-to-day effort satisfied his physical wants, and his mind was occupied with the host of friendly and malign spirits which filled the air, sea, land, ice and animals around him. In the long winters he found leisure to develop arts, skills, and an unaggressive social life. For all his hardships he seems to have been happy.

Exploitation

But his environment kept Arctic man at a Neolithic level of culture, far behind the advancing civilizations of the south, and much at their mercy. Wave after wave of invaders coming north to exploit new ground found Arctic man and his environment easy prey. Drawn by land hunger, following a nose for business, driven by conscience or political pressures, the newcomers brought tools, weapons, their own calendar of friendly and malign spirits, and a new and destructive way of life. Primitive man cajoled a living from the Arctic. Civilized man demanded both a living and a profit. Used to exploiting a richer environment, he took more than the Arctic could afford to give.

The impact of civilization has varied from place to place. Polar man in Eurasia suffered repeated invasion throughout mediaeval times; southerners bullied and taxed him, killed his reindeer, fought over his lands, and sent south a harvest of furs, reindeer skins, animal oils, whalebone – even the ivory of fossil mammoths. The natives absorbed their conquerors and carried on. Their final assimilation in the present century has been a relatively painless process. Herdsmen are settling to farming and market gardening, which technical innovation is now extending beyond the Arctic circle. Commercial fisheries replace the nomadic fishermen on the great estuaries of the northern rivers; the sons of reindeer herdsmen become mining engineers and accountants. Ruthless over-exploitation has turned to development—purposeful, bureaucratic, perhaps over-optimistic, but no longer cynically destructive.

Greenland and Iceland have supported precarious civilizations for over one thousand years. Greenland was thinly peopled by coastal Eskimos when the first Norwegian settlers arrived in AD 981. Iceland had no aboriginal population; its Irish and Scandinavian colonists began farming in the eighth century, and the two communities developed together as independent settlements trading with Scandinavia. Climatic deterioration in the fifteenth and sixteenth centuries brought hard times to both.

The Greenland farmers disappeared altogether from their strips of cultivated land in the south-west coast. Icelanders struggled on through famine, plague, and even volcanic eruption, to emerge once again as subsistence farmers when better climate returned in the late nineteenth century. Greenland was nursed back to economic health by Denmark; its modern inhabitants are of mixed Eskimo and Scandinavian stocks. Today both Iceland and Greenland are developing new prosperities based on efficient modern fishing industries. The Icelandic Government is balancing rival demands by industries, who want to flood parts of the country to develop hydro-electric power, and naturalists who see a threat to wildlife and the destruction of tourist amenities.

The Eskimos and Aleuts skirmished with Indians and other neighbours for centuries, but did not feel the full impact of civilization until whalers, sealers and trappers began to appear in their territories. Alaska and the Aleutian Islands were first colonized by Russian fur traders; later American sealers and whalers worked the Bering sea from the southern fur seal islands to the pack ice. Alaska was bought by the United States in 1867, and

has since been developed as a source of furs, timber, fish, precious metals and—most recently of all—mineral oil and petroleum. Aleuts disappeared gradually from the scene after harsh treatment by successive exploiters. Alaskan Eskimos and Indians of the mainland fared better, though their cultures have gradually been absorbed by an Arctic version of the American way of life. White settlers now outnumber them three or four to one, natives still predominating where traditional skills of hunting, fishing and simple Arctic survival are needed.

Eastern Eskimos and Indians have suffered demoralization. Their traditions are tattered, and

Right:
Commercial whaling. Cutting up a fin whale at a northern whaling station. Whalers were among the first to explore the Arctic and exploit its natural resources. Though still active, their industry has declined rapidly in recent years through over-exploitation; there are few large whales left to catch.

Whaling, one of the earliest Arctic industries, led to many voyages of exploration and discovery during the seventeenth and eighteenth centuries. The Greenland right whale was hunted almost to extinction in the nineteenth century.

their poverty and the bleakness of their environment have begun to show. Their battle was lost when, reasonably enough, they took guns and knives, fabrics and pans, medicines and liquor, to ease their day to day struggle. Paying for these simple luxuries in furs and other natural produce altered their role in Arctic economy; they too became exploiters, in an environment which cannot stand exploitation. Once poor but self-sufficient, they are now poor but dependent, misfits in the new civilization and aliens in their own country. Nowhere in the world do we see more clearly that, buying the benefits of civilization, man mortgages his future to pay for them.

Marine exploitation

Even more than Arctic lands, Arctic seas show the results of ruthless exploitation over several centuries. Hunted by man for nearly four hundred years, they have now lost almost completely their stocks of large mysticetes or whalebone whales (p. 58). Before steel and plastics became plentiful, whalebone was one of the few strong, springy materials known to man. Right whales – slow, easy-going plankton feeders of cold seas—have longer whalebone than other species, with individual plates measuring up to five metres. They also carry thick blubber and yield high-quality oil. Slow enough to harpoon from open boats, they did not sink after killing and were in every way the right whales to catch. The hunt for right whales began off northern Europe in the late sixteenth century, and ended when practically every northern right whale had yielded his whalebone for corset stays, umbrella ribs, brooms and other essentials.

From Norway the hunt shifted northward. Svalbard populations were wiped out by the mid-seventeenth century. Greenland stocks, protected by heavy pack ice, lasted until the late nineteenth century. A short-lived Bering Sea fishery collapsed from over-exploitation at about the same time. Humpbacked and grey whales suffered the same fate from the seventeenth century onward; the latter, which migrate annually along the American Pacific coast, are now staging a slow return under rigorous government protection.

Arctic sea hunters of early days have much to

BREUMMER

163

answer for. Apart from their ruthless whaling, they destroyed huge stocks of walruses and seals for oil, exterminated the great auk and Steller's sea cow (a large marine herbivore of the northern Pacific), Hudson Bay stocks of musk oxen, and many colonies of fur seals. They were not gentle with human populations, spreading more destruction and disease among Aleuts, Indians and Eskimos than will ever be accounted for. But they were hard men working for hard masters, and history will find excuses for them. It may be difficult to excuse, or even understand, their successors in the late nineteenth and twentieth centuries, who continued the destruction of whale stocks with greater technical efficiency, ceasing only when profit could no longer be made.

Steam power and the explosive harpoon made it possible for rorquals—the fast-moving blue, finback or fin, sei and minke whales—to be hunted from the late 1860s onward. As before, the hunt started off Norway, moved to Svalbard and Iceland, and finally involved floating factory ships in the polar oceans of both hemispheres. Up to the Second World War the annual catch was restricted only by the market for oil, and whales of all sizes and both sexes were taken. Later an international commission of eighteen nations met twice annually to determine, with the advice of scientists, an upper limit for the annual catch, to set restrictions on size of animals taken, protect mothers with young in attendance (but not other females), and divert catch effort from more popular to less popular species. The commission set itself the impossible task of controlling a profitable, highly capitalized, and fiercely competitive industry—one which did not want to be controlled, and could not work profitably at less than its peak of efficiency. Not surprisingly, the restrictions imposed by the commission were politically expedient but fell short of good husbandry. Now the world has lost practically all of its blue whales and most of its finbacks. Hunting pressure has fallen hard upon the smaller species, though their oil yield is poor and whalers have previously judged them barely worth the killing. Pressure falls even harder on remaining sperm whales, largest of the odontoceti or toothed whales, for they are large enough to show profitable hunting while stocks last.

Destruction of whalebone whales was offensive on many grounds, not least on grounds of common sense. They were a group of animals useful to man, which could have been cropped rather than quarried. Whalebone whales transform zooplank-

BREUMMER

Inukshuk. Eskimo markers overlooking a river route, Hudson Bay.

ton—a plentiful, reliable food resource which man cannot eat—into protein and oil which he can. No other animal, and no machine of man's devising, does the job so efficiently and cheaply. Our successors will marvel that civilized man destroyed, with single-minded determination and under international supervision, the creatures which unlocked for him the capacious food cupboards of polar seas.

Arctic animals and man

". . . the north can produce what the world needs . . . the coming Arctic boom is only a few decades away." So say the writers of a recent polar book,* and ecologists shudder at the implications. For centuries the Arctic kept primitive man in his place. It cannot cope at all with civilized man. Wielding unlimited power and energy, casting money before him and rubbish over his shoulder, modern man is marching north to mend his fortunes. He has already marched east, south and west; only the north remains open to him. Devils of his own making – call them national pride and progress, political neurosis and greed, or what you will – demand nothing less than the full, immediate exploitation of the world's resources for his benefit. From the Arctic, man is demanding oil to burn in his down-town traffic jams, copper and iron to fill his overflowing scrap-yards, poor tundra farmland

*Willy Ley and the editors of Life: see bibliography.

164

SWITHINBANK

New industry for North America—new hazard for the Arctic environment. Oil rig on the north Alaskan coast.

to replace the acres of good soil he has sunk under concrete in the south. However lunatic their demands, at whatever cost to the Arctic environment, his devils will get what they want.

While the maniac phase of expansion lasts, Arctic animals are at hazard. Over-cropping is a constant danger, as whales, caribou, great auks, musk-oxen, and a hundred species of warmer regions can testify. Used to the cornucopia of southern temperate ecology, man cannot accept the basic poverty and slow turnover of Arctic economy. All his dealings with Arctic wildlife degenerate rapidly into robbery. A danger even greater is pollution, for the simple polar ecosystems have fewer buffers to protect them against the invasion of foreign chemicals. Pesticide residues already poison the fats of polar sea birds and seals; oil slicks are commonplace on Arctic water even before the supertankers have begun their regular runs. Strontium-90 and cesium-137 from atmospheric fall-out are built into the tissues of Arctic lichens and the herbivores which feed on them. Reindeer and caribou have high levels of radioactive strontium in their bones, and all animals which feed on them are effected. Lapps in 1965 suffered the highest radiation exposure of any human population – from the reindeer meat which is their everyday diet.

Even the garbage of civilized man is a pollution problem on the tundra. There is a lot of it, for modern polar man brings half his world with him in expensive packaging. It will not rot in the cold

air, cannot be buried in the frozen ground, and no-one will pay to burn it or carry it back to where it came from. Every tundra settlement has its mountain of imported rubbish, its downwind litter of torn cartons, packing notes, circulars, unheeded instructions, and forms-in-triplicate which nobody filled in.

The Arctic is changing, and its animals face new problems and challenges. Will they survive? Is any part of the Arctic safe for them? The safest environment for any species is the one which changes least and most slowly. At present Eurasia holds a brighter future for Arctic animals than North America. The Soviet polar bear is part of a plan – a deliberate, sometimes a bumbling policy, to develop Eurasia for man. There is the thought and research of several decades behind the plan, and considerable respect for the environment. Progress is slow – probably slow enough for the Soviet bear to find and keep his place within the system.

The North American polar bear is heading for trouble. Though almost completely protected from simple bloodlust, he stands where the boom – when it comes – will hit hardest. At present there is little evidence of planning in North American policies, only the restless ebullience of two nations which, having exploited their continent to the edge of devastation, refuse to be halted by the tree line. When the Arctic booms, the American bear will be asked, politely but firmly, to move over and make room for some real-estate development. There may be nowhere for him to move to.

Bibliography

BAIRD, P D: *The Polar World*
Longmans, London, 1965

BÖCHER, T W; HALMEN K and JAKOBSEN K: *The Flora of Greenland*
Haase and Son, Copenhagen, 1968

BROOKS, C E P: *Climate through the Ages*
McGraw Hill, New York, 1949

COON, C. S: *The Living Races of Man*
Cape, London, 1966

DUNBAR, M. J: *Ecological Development in Polar Regions*
Prentice Hall, New Jersey, 1968

FREUCHEN, P; and SALOMONSEN, F: *The Arctic Year*
Cape, London, 1959

HOPKINS, D M (ed.) *The Bering Land Bridge*
Stanford University Press, 1967

KIMBLE, G H T and GOOD, D: (eds.) *Geography of the Northlands*
Wiley, New York, 1955

KING, J E: *Seals of the World*
British Museum, London, 1964

LEY,W and the editors of *Life* magazine: *The Poles*
Time–Life, New York, 1963

LINDROTH, C H: *The Faunal Connections Between Europe and North America*
Wiley, New York, 1957

MACKINTOSH, N A: *The Stocks of Whales*
Fishing News, London, 1965

MITCHELL, J M (ed.) *Causes of Climatic Change*,
Meteorological Monographs 8 (30), American Meteorological Society, Boston, 1968

POLUNIN, N: *Circumpolar Arctic Flora*
Oxford University Press, 1959

STEFFANSSON, V: *The Friendly Arctic*
Macmillan, London, 1921

SUSLOV, S P: *Physical Geography of Asiatic Russia*
W. H. Freeman, San Francisco and London, 1961

VOOUS, K. H: *Atlas of European Birds*
Nelson, London, 1960

Glossary

Arctic The north polar region, bounded by the tree line or the 10°C July isotherm.

Boreal forest The broad zone of coniferous forest of northern Canada and Eurasia.

Canadian Shield The ancient granite, granite-gneiss and sedimentary rocks, of pre-Cambrian age, which underlie northeastern Canada and the Canadian archipelago.

Floe Sheet of floating ice, often a detached part of a larger ice field.

Glacial period The period within an ice age (see pages 44 and 45) when persistent cold leads to the spread of large ice sheets over the land. The present ice age has included at least four glacial periods, separated by interglacial periods when the ice sheets dispersed or retreated in response to climatic changes.

Gulf Stream The oceanic current of warm water which crosses the north Atlantic Ocean from the Gulf of Mexico to northern Europe, bringing warm water and air to the Arctic in this sector; also called the North Atlantic Drift.

Interglacial period see **Glacial period**.

Interstadial see **Stadial**.

Isotherm A line drawn on a map connecting points on the earth which have the same mean temperature. The 10°C July isotherm connects points where the mean monthly temperature for July is 10°C.

Lead A crack of open water developing in an ice sheet; caused by tension due to tidal movements, currents, or other disturbance under the ice.

Moraine Rock debris transported and left behind by a glacier.

Periglacial Close to a glacier; periglacial soils are those which develop close to, and under the influence of, a glacial ice sheet.

Polynya A lake of open water in the middle of an ice sheet.

Scree A mass of detritus, generally broken rock, at the foot of a cliff; usually in the form of a sloping mound. Talus has the same meaning.

Subarctic Zone of the earth's surface bounded by the 10°C July isotherm to the north, in which mean temperatures do not exceed 10°C for more than four summer months, and the mean temperature of the coldest month is below freezing point.

Stadial The period within the glacial period during which ice sheets reach maximal spread; separated by interstadials, during which the sheets retreat slightly—though not so far as in an interglacial.

Talus See **scree**.

Tree line Poleward limit of tree growth; in the northern hemisphere, the line north of which trees do not develop full stature.

Tundra Treeless region; Arctic tundra lies north of the tree line, alpine tundra lies above the tree line on mountain slopes.

Index

Numbers in italics refer to illustrations.

A

Alaska
  development of, 162
  Northern, 27
Alcids, 84, *89*
  auk, great, *Alca impennis*, 89, 164
  auk, little, *see* dovekie
  auklets, 88–9
    least, *Aethia pusilla*, 89
    crested, *Aethia cristatella*, 89
    parakeet, *Cyclorrhynchus psittacuta*, 89
  dovekie (little auk), *Plautus alle*, 88, *89*
  guillemot (murre), 84, *84*, *85*, 86
    Black guillemot *Cepphus grylle*, *84*, 86
    Brunnich's, *Uria lomvia*, *84*, 86
    Common, *Uria aalge*, 84, *85*
    Pigeon guillemot *Cepphus columba*, 86
  murre, *see* guillemot
  murrelet, Kittlitz's, *Brachyrhamphus brevirostris*, 88–9
  puffin, *Fratercula arctica*, 86, *87*, 88, *88*
    horned, *Fratercula corniculata*, 88
    tufted, *Lunda cirrhata*, 88
  razorbill, *Alca torda*, 88
Aleut, 154, 162, 164
alligator, 44
amphibians, 33
Antarctica, 39, 40, 46, 74, 83
animals and plants, interaction
  between, 52
antelope
  North American prong-horn,
    Antilocapra americana, 47
  steppe, *Saiga tartarica*, 48
Arctic
  boom in, 164
  defined, 16
  ecology of, 31
  evolution of, 36, 41
  impact of civilization on, 161
Arctic animals
  adaptability, 32
  early, *46*
  on the tundra, 16
  origins of, 32
  success of warm-blooded animals, 33
  uniformity of, 32
Arctic basin, 39, 43
Arctic circle, 14
Arctic islands, 29
Arctic shorelines and seas, ecology of,
  54
Arctic-Subarctic boundary, 16, *16*, 20
auklet, *see* Alcids
aurochs, *Bos primigenius*, 48

B

badger, *Meles meles*, 48
bat,
  Hoary, *Lasiurus cinereus*, 136, *136*
  Red, *Lasiurus borealis*, 136, *136*
Bathurst Inlet, *26*
bathypelagic animals, 53
bear, 45, 165
  black, *Ursus americanus*, 124, *126*
  brown (Kodiak), *Ursus middendorff*,
    124, 126
  grizzly, *Ursus horribilis* 124, *125*
  land bears on the tundra, 124
  polar, *Thalarctos maritimus*, *18*, 20, 24
    27, 30, 31, 68–70, *70*, *71*; mating
    habits, 68–70; restrictions on killing,
    156
bearberries, *100*
beaver, 45
bee, bumble *Bombus sp.*, 109
beetles, 106, 109
beluga, *see* whale, white, 59, *61*
bethnic animals, 56
Beringia, 44, 45, 47, 48
birds, 33
  migrating, 20, 29, *29*, 31, 32, 56, 83,
    144, 145, 147, 150
bison, *Bison bison*, 46
  long-haired, 47
  plains, 48
blackfly (Simuliidae), 109
blowflies, *Calliphora sp.*, *Scatophaga sp.*,
  109
bluethroat, *Luscinia svecica*, 149
Brooks, C.E.P. 40, 42
bullheads (Cottidae), 56
bunting,
  Lapland, *Calcarius lapponicus*, 138
  snow, *Plectrophenax nivalis*, *138*, 139
butterflies, 106, *107*

C

Canada, northern, 27, 28
capelin, *Mallotus villosus*, *54*, 57
Carbo-Permian ice age, 31
carnivores, 44, 45, 46
  on the tundra, 124
caribou, *Rangifer tarandus*, 20, 29, 43,
  48, 95
  barren ground, *19*, *26*, *113*
  Greenland, *113*
  Peary, *112*
  effect of radioactive substances on, 165
char, *22*
  Arctic, *Salvelinus alpinus*, 57, *57*, 58
chipmunks (Sciuridae) 45
Chukchi, 154, *155*
clam, *54*, 56
coalfish (saith), *Merlangus carbonarius*, 57
cod, *Gadus sp.*, *54*, 57
continental drift, *40*
cormorant, 76
  common, *Phalacrocorax carbo*, 76
  green, *Phalacrocorax aristotelis*, 76
coyote, *Canis latrans*, 127
  on the tundra, 118, 127, *127*, *128*, *143*

underground burrows, 130
cranefly, Tipulidae, 106
crustaceans, 55
curlew, *Numenius arquata*, *146*

D

damselflies, 108, 109, *109*
deer, 45–8
  fallow, *Dama dama*, 46
  primitive, 47
  red, *Cervus elaphus*, 47
  roe, *Capreolus capreolus*, 47
diver (loon)
  black-throated, *Gavia arctica*, *90*, 91
  great northern, *Gavia immer*, *90*, 91
  red-throated, *Gavia stellata*, *90*, 91
  white-billed, *Gavia adamsii*, 91–2
dog, *33*, 157
  sledge, 158, *158*
dotterel, *Eudromias morinellus*, 149
dovekie, *see* Alcids
dowitcher, *Limnodromus griseus*, 149
dragonflies, *108*, 109
duck, of the tundra, 138, *145*
  Blue-winged teal, *Anas discors*, 144
  diving, 144
  gadwall, *Anas strepera*, 144
  green-winged teal, *Anas carolinensis*, 144
  mallard, *Anas platyrhynchos*, 144
  pintail, *Anas acuta*, 144
  teal, *Anas crecca*, 144
  wigeon, *Anas penelope*, 144
duck, sea, 92, *92*
  common scoter, *Melanitta nigra*, 92
  eider, *Somateria mollissima*, 92, *93*
  golden-eye, *Bucephala clangula*, 92
  goosander, *Mergus merganser*, 92
  harlequin, *Histrionicus histrionicus*, 92
  king eider, *somateria spectabalis*, 92
  oldsquaw, *Clangula hyemalis*, 92
  red-breasted merganser, *Mergus
    serrator*, 92
  spectacled eider, *Arctonetta fischeri*, 92
  Steller's eider, *Somateria stelleri*, 92
  velvet scoter, *Melanitta Fusca*, 92
dunlin, *Calidris alpina*, 149

E

eel-pout (Zoarcidae), 56
elephant, *Elephas sp.*, 46, 48
  grazing, 47
elk, *116*
  European, *see* Moose
  giant, *Megaceros sp.*, 46
  North American (wapiti), *Cervus
    canadensis*, 116
ermine, *see* weasel, short tailed
Eskimo, 27, *42*, 154
  clothing, 157, *158*
  in Greenland, 161–2, 164
  Mongoloid origins of, 154
  occupations of, *66*, 154, *154*, 156
    156, 157

Europe, Arctic, first settlers in, 154;
  Northern 22, 24
Eurasia, 15
  as a future safe environment, 165

F

fauna
  American, 45
  European mammal, 45
  mammal, 47
falcon
  Gyr falcon, *Falco rusticolus*, 150, *150*
  peregrine, *Falco peregrinus*, 150, *151*
fieldfare, *Turdus pilaris*, 149
Finns, 154
fish
  ability to withstand cold, 52
  Arctic bottom feeding, 57
  benthic, 53
  eggs of polar, 57
  larval, 55
  littoral or intertidal, 53
  movement of with plankton, 56
  sublittoral, 53
  variety and concentration of in
    subarctic waters, 57
fishery, Bering Sea
  decline of, 163
fishing
  development of commercial, 161
  exploitation of fishing grounds, 163
  restrictions on, 164
flowering plants and shrubs, 28, *28*, 31, 98
  bearberry, *100*, 105
  bilberry, 105
  bluebell, 105
  burr reed, 106
  campion, 105
  crowberry, 105
  heather, 105, *106*
  heaths, 105
  lupin, *106*
  oysterleaf, 106
  rhododendron, *106*
  rush, *105*
  saxifrage, 105, *106*
  sea chickweed, 106
  sedge, 105
  shinleaf, 105
  water buttercup, 106
forest-tundra zone, 16
fox
  Arctic, *Alopex lagopus*, 22, 29, 31, 48,
    124; on the tundra, 118, 124;
    variation in colour, 130, *130*
  red, *Vulpes fulva*, *32*, *131*
frog, *Rana sp.*, *30*, 33
fulmar, *see* petrel
Franz Josef Land, 31

G

gannet, northern, *Sula bassana*, *73*, 75–6
geese, 24, 29, 138, 144, *144*, 146
  barnacle, *Branta leucopsis*, 146
  bean, *Anser fabalis*, 147

blue, *Anser caerulescens*, 147
Brent, *Branta bernicla*, 146
pink-footed, *Anser brachyrhynchus*, 147
red-breasted, *Branta ruficollis*, 146
Ross's, *Anser rossi*, 147
snow, *Anser hyperboreus*, *29*, *144*, 147
white-fronted, *Anser albifrons*, 146
glacial periods (Alpine, Scandinavian,
  American), *44–5*, 46–9
gnats, *Tanypus sp.*, 109
goat, mountain, *Oreamnos americanus*,
  46, 48
godwit, 149
  bar-tailed, *Limosa lapponica*, 149
  black-tailed, *Limosa limosa*, *91*, *147*, 149
  Hudsonian, *Limosa hoemastica*, 149
Gondwanaland, 38–40
grasses, 16, 22, 27, 28, 29, 45, 48, 98, 99,
  104, 109, 114, 124, *146*, 147
  cotton grass, *105*
  lyme grass, 106
  mare's tail, 106
grass seed, 139
grazing animals, 46
great auk, *see* Alcids
Greenland, 20, 22, 161
guillemot, *see* Alcids
Gulf Stream, *see* North Atlantic Drift
Günz glacial period, 47
gull, 77, *77*
  black-headed, *Larus ridibundus*, *78*
  common (mew), *Larus canus*, 82–3
  glaucous, *Larus hyperboreus*, 78, *79*
  great black-backed, *Larus marinus*, 78,
    79, *104*
  herring, *Larus argentatus*, *79*
  Iceland, *Larus glaucoides*, 82, *83*
  ivory, *Pagophila eburnia*, 78
  kittiwake, *Rissa tridactyla*, 77, *80*
  lesser black-backed, *Larus fuscus*, *79*
  Ross's, *Rhodostethia rosea*, 78
  Sabine's, *Xima sabina*, 78
  Thayer's, *Larus thayeri*, 82

H

haddock, *Gadus aeglefinus* and *Sebastes
  marinus*, 57
halibut, polar, *Reinhardtius hippoglossoides*,
  57
hare, 45, 48, 136, 144
  Arctic, *Lepus arcticus*, 20, 29, 48, 123,
    123
  blue, *Lepus timidus*, 123
  on the tundra, 105, 118, 123
  snowshoe, *Lepus americanus*, *122*, 123
  tundra, *Lepus othus*, 123
heather, *106*
herring, *Clupea sp.*, 54
Hippopotamus, *Hippopotamus sp.*, 45, 48
horse, *Equus sp.*, 48
  browsing, 42, 45
  three-toed, 45
  single-toed, 45
Hudson Bay, *53*
Hudson Bay Company, 118
hyaenids (Hyaenidae), 45, 48

I

ice cap, *35*, 38
ice floes, *14*
ice,
  annual pack, 52
  fast, 52
  permanent, 36
  polar pack, *37*, 38, *39*, *51*, 52, *53*
Iceland, 20, 22, *22*, *49*, 161, 162
insectivores, 44
invertebrates, 52; on the tundra, 106–10
isotherm, 15, *24*

J

jaeger, 83, 84, 118; in northern and
  southern hemispheres, 84
  longtailed, *Stercorarius longicaudus*, 84,
    84
  pomarine, *Stercorarius pomarinus*, 84
  Arctic, *Stercorarius parasiticus*, *82*, 83

K

kittiwake, *see* gull
Kolyma river, *14*
Knot, *Calidris canutus*, 149
Köppen, Wladamir, 14, 16

L

Labrador, *26*
Lapps, *160*, 165
lark, Horned, *Eremophila alpestris*, 149
Laurasia, 38, *39*
lemming, 22, 29, 31, 33, 45, 48, 105,
  118, 135, 136, 144
  bog, *Synaptonyx borealis*, *118*, 120
  brown, *Lemmus trimucronatus*, *118*, 120
  collared, *32*, 48, 120
    Arctic, *Dicrostonyx torquatus*, *118*, 120
    Greenland, *Dicrostonyx groenlandicus*,
      118, 120
    Hudson Bay, *Dicrostonyx hudsonius*,
      118, 120
  Norway, *Lemmus lemmus*, *118*, 120
  on the tundra, 118
  Siberian, *Lemmus sibiricus*, *118*, 120
  tunnels, 120
Lena river, *41*
lichens, *38*, *100*, *104*
ling, *Molva molva*, 57
lion, *Felis sp.*, 45, 48
lizard, 33
loon, *see* divers
Lupins, Arctic, *106*
lynx, 45–6, 48
  Canadian, *Lynx canadensis*, 136, *137*
  Eurasian, *Lynx lynx*, 136
  on the tundra, 120

M

magpie, *Pica pica, 143*
mammal
 Eurasian, 45
 terrestrial, 33
 Pleistocene, 48
mammoth, *Mammuthus sp.,* 45, *46*
 woolly, *47*
man
 first appearance of, 48; in the Arctic,
 153, 154, 161, 164; similarity of
 cultures in the Arctic, 157
marmot, hoary, *Marmota caligata, 120*
 on the tundra, 118, 123
marsupial, 44
marten, 48, *134*
 North American, *Martes americana,*
 133, *134*
 pine, *Martes martes,* 133, *134*
martin, sand, *Riparia riparia, 147,* 149
mastodon, *Mammut sp.,* 42, 45, 46, *46*
merlin, *Falco columbarius,* 150
Mesozoic era, 38–9, 42
mice, meadow, *Microtus sp.,* 45
midges (Chironomidae), 109
Milankovich, M., 42
Mindel-Elster-Kansan glacial period,
 47
mink,
 European, *Mustela lutreola,* 133, *133*
 North American, *Mustela vison,* 133,
 133
molluscs, 56
moose, *Alces alces, 16,* 47, 48, *115*
mosquito, *Aedes sp., 108,* 109
moss, 28, *39, 97*
murrelet, *see* Alcids
musk ox *Ovibos moschatus, 15,* 20, *27,*
 29, 31, 48
 exploitation of, 164; grazing herds,
 110, 111, *112;* on the tundra, 99;
 recovery of the species, 112; retreat
 northward, 112; woodland,
 Bootherium sp., 48
muskrat, *Ondatra zibethica,* 118, *120,*
 122, 123
Mustelidae, 130, 133, 135

N

Neanderthal man, 48
Nebraskan glacial period, 47
nekton, 53
nitrogen, 106, enrichment 74, 96
North American Indian, 154, 164
North Atlantic Drift, 18, 22, *23,* 24,
 31, 38
North Pole, 40, 44
Northwest Passage, 29

O

oil, resources of in Arctic, *165*
otter, European, *Lutra lutra,* 130, 133,
 132

North American, *Lutra canadensis,* 130,
 133, *133*
 sea, *Enhydra lutris,* 70
owl
 great grey, *Strix nebulosa, 148,* 149
 short eared, *Asio flammeus,* 149
 snowy, *Nyctea scandiaca,* 118, *127,* 143,
 143, 149
oystercatcher, *Haematopus ostralagus,*
 145, 147

P

Palaeozoic, 39
passerines, 149
pelagic animals, 53
permafrost, *20, 23,* 27, *28*
Permian rock, 40
petrel, 74, *74*
 Atlantic storm, *Hydrobates pelagicus,*
 74, 75
 fulmar, *Fulmarus glacialis, 74,* 74–5,
 75
 Leach's storm petrel, *Oceanodroma*
 leucorhoa, 74, 75
 shearwater, Manx, *Puffinus puffinus,*
 74, 75
 slender-billed, *Puffinus tenuirostris, 74,*
 75
 sooty, *Puffinus gravis, 74,* 75
Permo-Carboniferous ice age, 36
phalaropes, 76–7, 144
 as mid-ocean feeders, 76
 grey, *Phalaropus fularicus,* 76
 rednecked, *Phalaropus lobatus, 76, 76,*
 105
phytoplankton, 20, 52–3, *54*
photosynthesis, 56
pig, *Sus sp.,* 48
pika, *Ochotona princeps,* 45, *119,* 123
pipit, meadow, *Anthus pratensis, 148,*
 149
 red-throated, *Anthus cervinus,* 149
 rock, *Anthus spinoletta,* 149
plankton, 20, 52, 56
Pleistocene era, 20, 27, 28, 29, 30, *36,*
 38, 42, 45, *46*
Pliocene, 44, 45
plover, 147, 149
 blackbellied (grey), *Squatarola*
 squatarola, 91, 147
 golden, American and Asian,
 Pluvialis dominica, 91, 146, 147
 European, *Pluvialis apricaria, 91,* 147
 semipalmated (ringed), *Charadrius*
 niaticula, 91, 149
polar basin, 18, 20, 22, 32, 38, 48
poles, magnetic and geographic, 39–40
polynyas, 52
poppy, Arctic, *15,* 105
porcupine, *Erethizon dorsatum, 120, 122,*
 123
primates, 44
ptarmigan, 29, 144
 on the tundra, 99, 118
 rock, *Lagopus mutus,* 139, *140,* 143, 150
 willow, *Lagopus lagopus,* 140, *141,* 143,
 150
puffin, *see* Alcids

R

rabbit, *Oryctolagus sp.,* 45
raven, *Corvus corax,* 143, *143*
razorbill, *see* Alcids
redpolls, *Acanthis flammea,* 139, *139*
redwing, *Agelaius phoeniceus,* 149
reindeer, *Rangifer tarandus, 23,* 24, 31,
 114, 157, *161*
 as pack animal, 157
 effect of radioactive substances on,
 165
 European, 48
reindeer moss, *Cladonia sp.,* 22, 28, 29,
 100
reptiles, 33
rhinoceros, woolly, *Coelodonta sp.,* 45,
 46, 47, 48
Riss-Saale-Illinoian glacial period, 48
rodents, 45, 48
ruff, *Philomachus pugnax, 91,* 149
ruminants, 45

S

saith, *Pollachius virens,* 57
Samoyeds, 154
sanderling, *Calidris alba,* 149
sandpiper, purple, *Calidris martima,* 149
sculpin, *Cottus sp.,* 56
sea birds, 57
 effect of pesticides on, 165
 feeding habits of, 74
sea cow, Steller's, 164
sea lion, 44
 Steller's, *Eumetopias jubatus, 62,* 64
sea-snail (Fish: Liparidae), 57
seal, 20, 24, 27, 30–1, 33, 44, 52, 58
 bottom-feeding species, 56
 breathing holes of, 66
 characteristic harem groups, 64
 effect of pesticides on, 165
 evolution of, 62
 exploitation of by hunters, 164
 interlocking teeth of, 53
 map showing distribution of, *64*
 mating habits, 64
 sonar techniques of, 56
 Arctic species of, 56
 three families of, 59, 62, 64
 bearded, *Erignathus barbatus,* 66
 common, *Phoca vitulina,* 67–8, *69*
 grey, *Halichoerus grypus,* 67–8, *68–9*
 harp, *Pagophilus groenlandicus, 39,* 53,
 66–7, *68*
 hooded, *Crystophora crystata,* 66–7, *67*
 Otariida in the Bering Sea, 64
 phocid, *64–5,* 65–6
 Pribilov fur seal, *Callorhinus ursinus,*
 63, 64–5
 ribbon, *Histriophoca fasciata,* 68
 ringed, *Pusa hispida,* 66, *67*
sealskin, 157
shag, *see* green cormorant
shark,
 Greenland, *Somniosus microcephalus, 54,*
 57, *59*

shearwaters, *see* petrel
sheep, 46, 48
  big horn, *Ovis canadensis*, 116
  Dall, *Ovis dalli*, 116, *116–7*
  on the tundra, 105
shore birds, *see* wader
shrew, 45, 135
  Arctic, *Sorex arcticus*, 136, *136*
  common, *Sorex araneus*, 136, *136*
  dusky, *Sorex obscurans*, 136, *136*
  masked, *Sorex cinereus*, 136, *136*
  Siberian, *Sorex sibiricus*, 136, *136*
shrub ox, *Eluceratherium sp.*, 45
Shrubs, *see* flowering plants
Siberia, 38
Siberian Arctic, 24, 27
skua, *Catharacta skua*, 83–4; great skua,
  82, 83–4
snake, 33
South Pole, 39, 40
Soviet Arctic islands, 31
sponge, *54*, 56
springtails (Collembolae), 109
squid, 53
squirrel, 45
  Arctic ground, *Citellus undulatus*, *29*,
  33, 48, 121, 123
  elaborate tunnels of, 123
starfish, *54*
stint,
  little, *Calidris minuta*, *147*, 149
  Temminck's, *Calidris temmincki*, 149
stoat, *Mustela erminea*, 24, 135
Strontium-90, 165
Subarctic, 16, 18
subarctic water, 18–20
Svalbard, *14–5*, 30–1, *41*
sun
  position of in the Arctic, 36, *37*
swan, 138
  Bewick's *Cygnus bewickii*, *142*, 144
  whistling, *Cygnus columbianus*, *142*, 144

T

Taiga, 16
tapir, *Tapirus sp.*, 42, 45
tern, Arctic, *Sterna paradisea*, *31*, 83,
  83, 103
Tertiary era, 40, 42, 43, 45
tiger, sabre-toothed, *Smilodon sp.*, 42,
  45, 46, *46*
tree line, 14, 15–6, *16, 17, 36*
trees, 14, 15, 16, 42
  alder, 27, 28, 29, 44, 49, 99
  birch, 22, 29, 44, 99
  elm, 44, 49
  fir, 44
  hazel, 44
  hemlock, 44
  hornbeam, 44
  larch, 27, 44, 45
  lime, 49
  oak, 44, 49
  spruce, 22, 27, *36*, 44, 45
  willow, 22, 27, 29, 31, 44, 98, 99, 102,
  104, 105
  world's northernmost, 26–7

tundra, 31–32
  ahumic soil of, 96, *106*; Arctic tundra,
  102; Arctic animals on, 16; Arctic
  brown soil, 97–8, 109; barren ground,
  15; birds on, 138, 144; calcium-rich
  soil, 98; carnivores on, 124; coastal
  tundra, *18*; cycles of herbivorous
  animals on, 118; cycles of predator
  species on, 118, 120; deer (Rangifer
  terandus), 113–6, *114*; development
  of Arctic soil, 96; dwarf willow on,
  104; effect of vegetation on, 96;
  forest tundra, 16; garbage on, 165;
  insects of, 106; insectivores on, 136;
  lakes and ponds on, *21*, 97, 110;
  miniature forests on, 99; mosquito
  and blackfly on, *28, 108*, 109;
  mountain tundra, 116; moose and
  elk on, 116; plant communities on,
  98–106, *99*; polar desert soil of, 98;
  sedges and reindeer moss on, 105;
  subarctic animals on, 16; variation of
  vegetation on, 118; wetland
  communities on, 105–6
Tungus, 154
turnstone, *Arenaria interpres*, 147
turtle, 44

V

vole, 33, 38, 135, 136, 144
  Alaska, *Microtus miurus*, *119*, 120
  insular, *Microtus abbreviatus*, 120
  meadow, *Microtus pennsylvanicus*, 48,
  119, 120
  on the tundra, 99, 118, 120
  pine, *Pitymys pinetorum*, 45
  tundra redback, *Clethrionomys
  rutilus*, *119*, 120

W

wader, 24, *91*, 138, 147
  breeding ground of, *105*
  on the tundra, 99
wagtail, 149
  citrine, *Motacilla citreola*, 149
  white, *Motacilla alba*, 149
walrus, *Odobenus rosmarus*
  as a bottom feeder, 56
  bull, 30
  exploitation of by hunters, *154*, 164
  functional heads of, 62, *63*
  food, 62
  unmated males, 62
wapiti, *see* elk, North American
warble fly, reindeer, *Oedemagena
tarandi*, *108*, 110
weasel, 24, 130, 135
  European, *Mustella nivalis*, 134, *135*
  least, *Mustela rixosa*, 135
  short-tailed (ermine), *Mustela
  erminea*, *134*, 135, *135*
West Greenland ice cap, *21*
whale, 20, 33, 52, 53, 54, 58, *60–1*
  Atlantic right, *Eubalakna glacialis*, 163
  blue, *Sibbaldus musculus*, 58, *60*, 164

  bottlenose, *Tursiops truncatus*, 58, *61*
  Fin (Finback), *Balaenoptera physalus*,
  58, *60*, *163*, 164
  Greenland right, *Balaena mysticetus*,
  162, 163
  grey, *Eschrichtius glaucus*, 58, *61*, 163
  humpback, *Megaptera novaeangliae*, 58,
  61, 163
  killer, *Orcinus orca*, 58, 59, *61*
  minke (piked), *Balaenoptera
  acutorostrata*, 58, *60*, 164
  Mysticetes, 58
  narwhal, *Monodon monoceros*, 59, *61*
  Odontocetes (toothed), 58
  Pacific right, *Eubalaena sieboldi*, *61*
  rorqual group, 58, 59, *61*, 164
  sei, *Balaenoptera borealis*, 58, *60*, 164
  sperm, *Physeter catodon*, 58, *61*, 164
  white (beluga), *Delphinapterus leucas*,
  59, *61*
  decline of, 163–4; movement of with
  plankton, 56; use of baleen, 55, 58,
  60–1, 163
whaling, *162–3*, 164
wheatear, Greenland, *Oenanthe
oenanthe*, 148
wolf, grey, *Canis lupus*, 24, 47, 48, *126*,
  127
  movement of with large mammals, 127
  on the tundra, 118, 127
  variation of colour in the Arctic, 127,
  127
wolverine, *Gulo luscus*, 45, 131
  fur, 157
Würm-Warthe-Weischel-Wisconsin
  glacial period, 48, 112

Y

Yakuts, 154
yellowhammer, *Emberiza citrinella*, *149*
Yenesi-Ostyaks, 154
Yukaghirs, 154

Z

zooplankton, 53, 55, 56